# HIGH-SPEED PHOTODIODES IN STANDARD CMOS TECHNOLOGY

# THE INTERNATIONAL SERIES IN ENGINEERING AND COMPUTER SCIENCE

## ANALOG CIRCUITS AND SIGNAL PROCESSING
*Consulting Editor*: Mohammed Ismail. *Ohio State University*

# HIGH-SPEED PHOTODIODES IN STANDARD CMOS TECHNOLOGY

*by*

**Saša Radovanović**

*National Semiconductor,*
*The Netherlands*

**Anne-Johan Annema**

*University of Twente, Enschede,*
*The Netherlands*

and

**Bram Nauta**

*University of Twente, Enschede,*
*The Netherlands*

A C.I.P. Catalogue record for this book is available from the Library of Congress.

ISBN-13 978-1-4419-3944-9
ISBN-10 0-387-28592-X (e-book)
ISBN-13 978-0-387-28592-4 (e-book)

Published by Springer,
P.O. Box 17, 3300 AA Dordrecht, The Netherlands.

*www.springer.com*

*Printed on acid-free paper*

# Contents

# CHAPTER 1

---

# Introduction

---

In the last decades, the speed of microprocessors has been increasing exponentially with time and will continue to do so for at least another decade. However, the local computing power of the microprocessor alone does not determine the overall speed of a system. Equally important to the processor's bare computing power is the speed at which data can be distributed to and from the processor. That means that the speed of the data-input and -output channel must keep pace with the processor's computing power. For the future it is expected that these data-communication channels will become the speed-bottle-neck for the whole system.

For short and medium distance (centimeters up to hundreds of meters) the data communication channels are usually implemented as wired electrical connections. However, at high speeds major problems occur: poor impedance matching results in distorted signals, signal losses due to the skin-effect, significant Electro-Magnetic noise is generated which degrades the system performances. In order to increase the data-rate in short-haul communication, the electrical wires can be replaced by optical fibers. The main focus is given at the receiver side; the objective was to design a low-cost Gb/s receiver that can be easily integrated with the rest of electronic circuitry.

The electronics for the long distance channels is typically realized with

1

expensive exotic technologies: Gallium-Arsenide High-Electron-Mobility-Transistors (GaAs HEMT) [1, 2], and Indium-Phosphide Hetero-Junction-Bipolar-Transistors (InP HBT) [3, 4]. The maximum bit-rate for these systems is around 100 Gb/s per channel. The first reason for adequacy of these expensive blocks is long distance links: the cost per length of the fiber is low. The second reason for the efficiency of this solution is that a large number of users share the links: the cost per user is low.

For medium and short distances however, as well as for a small number of users per link (fiber-to-the-home or fiber-to-the-desk) the optical receivers and transmitters should not be expensive. Because of the low cost requirement on the receiver (electronics), the complete optical detector should preferably be fully implementable in today's mainstream technology: CMOS. These receiver chips (inside microprocessors for example) have integrated light-sensors and thus they are cheap and do not have wire-speed limitations. The result could then be a low-cost and high-speed fully integrated optical data communication system for distances ranging from chip-to-chip (cm range) up to up to hundreds of meters, typical for LAN environments.

## 1.1   Outline

This book consists of 8 chapters. The goal is to design monolitically integrated optical receiver in straightforward CMOS technology, for short-haul optical communication and bit-rates up to a few Gb/s.

The second chapter gives a short introduction into optical interconnections. The advantages and disadvantages of the optical communication system in comparison with straightforward wired (electrical) communication channels are discussed. The three key building blocks for optical communication system, light sources, optical fiber, and light detectors are also discussed in chapter 2.

Chapter 3 presents a detailed analysis of the time and frequency responses of photodiodes in CMOS technology for $\lambda$=850 nm light. Physical processes inside a photodiode are thoroughly investigated using one particular demonstration CMOS technology: a standard 0.18 µm CMOS. The extention of the results to other CMOS technologies is also presented. For every high-speed photodetector there are two main parameters that define their figure-of-merit: responsivity and bandwidth. The bandwidth is the main limiting factor for Gb/s optical detection. There are actually two in nature different bandwidths of the photodiode: intrinsic (physical) and extrinsic (electrical) bandwidth. The first is inversely related to the time that excess carriers need to reach junctions and thus, to be detected at the output terminal. The second bandwidth is related to diode capacitance and the input impedance of the subsequent transimpedance amplifier. By approximation, the total bandwidth is the lowest between these two. These bandwidths will be separately analyzed in detail in chapter 3.

The intrinsic bandwidth of photodiodes in standard CMOS for $\lambda$=850 nm is typically in the low MHz range; this is two orders of magnitude too low for Gb/s data-rate applications. Chapter 4 presents a solution to boost the bitrate to over 3 Gb/s in standard CMOS technology without sacrificing diode responsivity. At the moment of writing this book this speed figure is over a factor 4 higher than other state-of-the-art solutions. This is achieved by using an inherently robust analog equalizer;complex adaptive algorithms are not required. The proposed configuration is robust against spread and temperature variations. Using this approach, 3 Gb/s data-rate for $\lambda$=850 nm and 0.18 μm CMOS technology with bit error rate BER=$10^{-11}$ at input optical power of $P_{\text{in}} = 25\mu W$, is demonstrated.

For very low wavelength $\lambda$=400 nm (blue light), the light penetration depth in silicon is very small (0.2 μm). Chapter 5 shows that then excess carriers are generated close to junctions which results in high bandwidths (hundreds of MHz up to a few GHz range).

Chapter 6 investigates polysilicon photodiodes designed using NMOS and PMOS gates. The measured bandwidth of the poly photodiode was 6 GHz, which figure was limited by the measurement equipment. However, the quantum efficiency of poly photodiodes is low (<8 %) due to the very small light sensitive volume. This active area is limited by a narrow depletion region and its depth by the technology.

Chapter 7 presents a generalization of the results in earlier chapters to photodiodes in any CMOS technology and operating on any sensible wavelength, from $\lambda$=400 nm to $\lambda$=850 nm. Also a generalization of the use of the analog equalization (introduced in chapter 4) to increase the operation frequency is presented.

Chapter 8 summarizes the most important conclusions in the book.

# Bibliography

[1] J. Choi, B.J. Sheu, Chen: "A monolithic GaAs receiver for optical interconnect systems O.T.-C.", *IEEE Journal of Solid-State Circuits*, Volume: 29, Issue: 3, March 1994, pp.328-331.

[2] C. Takano, K. Tanaka, A. Okubora; J. Kasahara: "Monolithic integration of 5-Gb/s optical receiver block for short distance communication", *IEEE Journal of Solid-State Circuits*, Volume: 27, Issue: 10, Oct. 1992, pp.1431-1433.

[3] M. Bitter, R. Bauknecht, W. Hunziker, H. Melchior: "Monolithic InGaAs-InP p-i-n/HBT 40-Gb/s optical receiver module", *Photonics Technology Letters, IEEE*, Volume: 12, Issue: 1, Jan. 2000, pp.74-76.

[4] H.-G. Bach, A. Beling, G.C. Mekonnen, W. Schlaak: "Design and fabrication of 60-Gb/s InP-based monolithic photoreceiver OEICs and modules", *IEEE Journal of Selected Topics in Quantum Electronics*, Volume: 8, Issue: 6, Nov.-Dec. 2002, pp.1445 - 1450.

CHAPTER **2**

# Short range optical interconnection

## 2.1 Why optical interconncction?

For nearly forty years scientists are using light to "talk" over distance. The birth of optical communications occurred in the 1970s with two key technology breakthroughs. The first was the invention of the semiconductor laser in 1962 [1]. The second breakthrough happened in September 1970, when a glass fiber with an attenuation of less than 20 dB/km was developed [2, 3]. With the development of optical fibers with an attenuation of 20 dB/km, the threshold to make fiber optics a viable technology for telecommunications was crossed. The first field deployments of fiber communication systems used Multimode Fibers (MMFs) with lasers operating in the 850 nm wavelength band. These systems could transmit several kilometers with optical losses in the range of 2 to 3 dB/km. The total available bandwidth of standard optical fibers is enormous; it is about 20 THz. A second generation of lasers operating at 1310 nm enabled transmission in the second window of the optical fiber where the optical loss is about 0.5 dB/km in a Single-Mode-Fiber (SMF). In the 1980s, telecom carriers started replacing all their MMFs operating at 850 nm. Another wavelength window around 1550 nm was developed where a standard SMF has its minimum optical loss of about 0.22 dB/km.

From this small history of fibers it can be concluded that the main research focus was on long-distance communication. Chapter 1 described that the electronics for the long distance channels is typically realized with expensive exotic technologies such as GaAs or InP. The bit-rate for these systems is large, around 100 Gb/s per channel, with low cost per length of the fiber and for a large number of users.

Replacing electrical wires with optical fibers for short distances for a small number of users is still challenging. The goal is to have low cost but high (Gb/s) bit-rates of the system. However, the important question is should we use light (fibers) to directly connect silicon chips and why?

A large study about this issue is published in the literature and some of the results will be briefly presented further in this chapter. In [4, 5, 6], Miller tried to stress the practical benefits of optical interconnects and drawbacks of electrical systems for high-speed communication (>10 GHz). His approach was to analyze the similarities and differences in optical and electrical systems, which will be briefly investigated in the following subsections.

### 2.1.1   Electrical and Optical Interconnection - Similarities

At the most basic level, optical and electrical physics are very closely linked. In practice, in both the electrical and optical case, it is the electromagnetic wave that carries a signal through a medium (see figure 2.1).

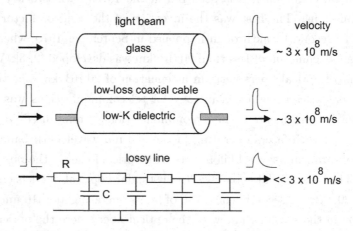

Figure 2.1: Types of optical and electrical propagation and their velocity. One possible model of the lossy line is presented.

It is important to stress that in high-speed communication, it is not electrons that carry the signals in wires or coaxial cables; actually the signal is carried by electromagnetic wave [4]. It is also good to note that signals in wires propagate at the velocity of light (or somewhat lower than light velocity if coaxial cables are filled with a dielectric). Hence it is generally incorrect to say that signals propagate faster in optics. In fact, signals typically travel slightly slower in optical fibers than they do in coaxial cables because the dielectric used in cables has a lower dielectric constant than glass.

In case of electrical interconnection lines on chips, the signals do move at a lower speed, but this speed is determined by the overall resistance (R) and capacitance (C) of the interconnect line [7].

## 2.1.2 Electrical and Optical Interconnection - Differences

Apart from large similarities, there are important basic differences between optical and electrical physics. The most important one is the higher (carrier) frequency and the corresponding large photon energy. The higher carrier frequency (shorter wavelength, typically in 1 µm range) allows us to use optical fibers to send optical signals without high loss [8]. There are small "wavelength windows" where the loss in the fibers (both singlemode and multimode) is small (<1 dB/km). The dispersion in singlemode and multimode fibers used in short distance communication is small too. In this way it is possible to avoid the major loss phenomena that in general limits the capacity of electrical interconnects on high frequencies: signal and clock distortion and attenuation.

The optical generation and detection for interconnection is in principle quantum mechanical (e.g., counting photons). This is in contrast to a classical source/detection of voltages and currents; for example, detection of light in practice involves counting photons, not measuring electric field amplitudes. Two practical consequences are that all optical interconnections provide voltage isolation (used in opto-isolators), and optics can offer lower powers for interconnects: it can solve the problem of matching high-impedance low-power devices to the low impedance (and/or higher capacitance) of electromagnetic propagation. With optical interconnection, there are no inductive voltage drops on input/output pins and wires that come for free in electrical interconnections.

A signal propagating down an electrical line may start with sharply rising and falling "edges". However, these edges will gradually decrease because of the loss-related distortion and dispersion, as illustrated in figure 2.1. This "softening" of the edges makes precise extraction of timing information more difficult. For the same communication distances, optical systems have relatively little problem with such variations. The dispersion and loss in optical fibers are typically smaller than in electrical wires, which is explained in section 2.3.2. Hence optic interconnect becomes increasingly attractive at high bit rates but also in higher interconnect densities (e.g., high density edge connectors for boards, or even very high density connections of chips), and arguments for optics become increasingly strong as the number of lines on the board increases. However, the disadvantage of optics is in the systems with optical connectors, because the connector size is much larger than the fiber diameter.

Optics also offers several additional opportunities that have essentially no practical analogy in the electrical case, including use of short pulses for improved interconnect performance [9]. A very important advantage of optical fibers is that they can be deployed in environments with large electromagnetic interference (EMI) and radio-frequency interference (RFI), such as airports, factories, military bases etc. In total, the advantages of optical interconnection in comparison with the straightforward electrical connection are summarized below, [4]:

- Immune to noise (electromagnetic interference and radio-frequency interference)

- Signal Security (difficult to tap)

- Nonconductive (does not radiate signals) - electrical isolation

- No common ground required

- Freedom from short circuit and sparks

- No inductive voltage drops on pins and wires

- Reduced size and weight cables (but not connectors)

- Ability to have 2-D interconnects directly out of the area of the chip rather than from the edge

- Resistant to radiation and corrosion

- Less restrictive in harsh environments

- Low per-channel cost [2]

• Lower installation cost in future (Wavelength Division Multiplexing [10])

Despite the many advantages of fiber optic systems, there are some disadvantages. Because of the relative newness of the technology, fiber optic components are still expensive even though the prices decrease dramatically in the last couple of years. Fiber optic transmitters (but not the receivers[1]) are still relatively expensive compared to electrical interfaces. The lack of standardization in the industry has also limited the acceptance of fiber optics. Many industries are more comfortable with the use of electrical systems and are reluctant to switch to fiber optics. However, the huge bandwidth advantage of the optical interconnection will probably force industry to move towards optic interconnect. Note that even with dominant optical interconnect, the on-chip signal processing remains electrical: an electrical-optical optical interface will always be required and probably the total speed in the system will be limited by the electronics.

## 2.2 Characteristics of light

The operation of optical communication and optical fibers depend on basic principles of optics and the interaction of light with matter. From a physical standpoint, light can be seen either as electromagnetic waves or as photons. Both view points are valid and valuable, but the simplest view for a fiber transmission is to consider light as rays travelling in straight lines and for a light detection to see the light as a number of incident photons on the photodetector surface.

Light is only a small part of the electromagnetic (EM) spectrum. The difference in radiation in different parts of EM spectrum is a quantity that can be measured: length of wave/frequency of EM-field and energy of photons. In some parts of the spectrum, frequency is used the most; in others wavelengths and photon energies are. In figure 2.2 the EM spectrum is presented with typical applications in certain spectral ranges.

In the optical world the most commonly used light quantity is wavelength, measured in micrometers or nanometers. It is inversely proportional to frequency $f$ and proportional to the speed of light $c$:

$$\lambda = \frac{c}{f} \tag{2.1}$$

---

[1]A 3 Gb/s data-rate optical receiver in inexpensive CMOS technology is presented in chapter 4.

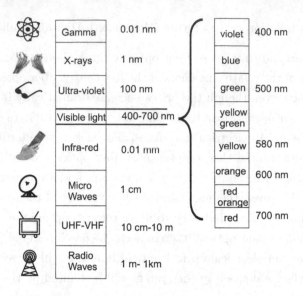

Figure 2.2: The electromagnetic spectrum.

## 2.3   Optical fiber types

Optical fibers are characterized in general by the number of modes that propagate along the fiber. Basically, there are two types of fibers: single-mode fibers and multi-mode fibers. The basic structural difference is the different core size.

### 2.3.1   Single-mode fibers

Single-mode fibers have lower signal loss and higher information capacity (bandwidth) than multimode fibers. They are capable of transferring higher amounts of data due to low fiber dispersion[2]. A cross section of a single mode fiber is shown in figure 2.3; this type of fiber is mainly used for long-haul optical communication because of low typical loss (typically lower than 0.2 dB/km).

### 2.3.2   Multimode fibers

As the name implies, multimode fibers propagate more than one mode; this is illustrated in figure 2.4. The number of modes, $M_n$, depends on the core size and numerical aperture (NA) and can be approximated by:

---

[2]Basically, dispersion is the spreading of light as light propagates along a fiber. This causes intersymbol interference i.e. an incorrect bit detection at the fiber's output.

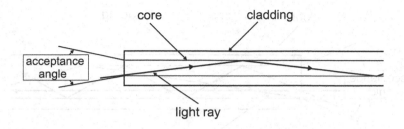

Figure 2.3: Single-mode optical fiber (small core diameter)

$$M_n = \frac{V^2}{2} \quad \text{and} \quad \frac{V^2}{4} \qquad (2.2)$$

for step index fiber and gradient index fiber, respectively. $V$ is known as the normalized frequency, or the V-number, which relates the fiber size, the refractive index, and the wavelength. The V-number is:

$$V = \left[\frac{2\pi a}{\lambda}\right] \times NA \qquad (2.3)$$

NA is closely related to the acceptance angle and it is approximately [8]:

$$NA = \sqrt{n_0^2 - n_1^2} \approx n_0 \sin \Theta_c \qquad (2.4)$$

where $n_0$ and $n_1$ are refractive index of the core and cladding respectively, and $\Theta_c$ is the confinement angle in the fiber core. As the core size and NA increase, the number of modes increases. Typical values of fiber core size and NA are 50 µm to 100 µm and 0.20 to 0.29 respectively.

A large core size and a higher NA have several advantages. Light is launched into a multimode fiber with more ease. Higher NA and larger core size make it easier to make fiber connections: during fiber splicing, core-to-core alignment becomes less critical. Another advantage is that multimode fibers permit the use of light-emitting diodes (LEDs). Single mode fibers typically must use laser diodes due to their small diameter ($< 10$ µm). LEDs are cheaper, less complex, and last longer and they are preferred for a large number of applications [8].

Nevertheless, multimode fibers have some disadvantages. As the number of modes increases, the effect of modal dispersion increases. Modal dispersion

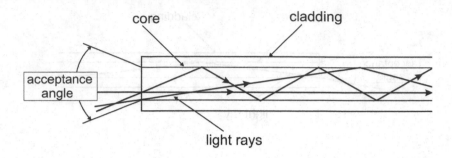

Figure 2.4: Multimode-mode optical fiber with multiple light rays. The angles of the light rays are refracted at the air/fiber interface according to Snell's law.

(intermodal dispersion) is important because, as the pulses spread, they can overlap and interfere with each other, limiting data transmission speed. Typical dispersion values for fiber are measured in nanoseconds per kilometer of fiber. These can be translated into an analog bandwidth limit in the transmission.

For instance, if one ray travels straight through a multimode fiber and another bounce back-and-forth at the acceptance angle $\Theta_c$ through the same fiber, the second ray would travel further for:

$$l_1 = l \left( \frac{1}{\cos \Theta_c} - 1 \right) \qquad [\text{m}] \tag{2.5}$$

where $l$ is the length of the multimode fiber. The ray that goes down the center of the fiber with speed $v$ will reach the output $\tau_r$ seconds before the the ray that bounces at the acceptance angle:

$$\tau_r \approx \frac{l_1}{v} \left( \frac{1}{\cos \Theta_c} - 1 \right) \tag{2.6}$$

Thus, an instantaneous pulse at the start will spread out $\tau_r$ seconds at the end. The analog bandwidth of the multimode fiber is inversely proportional to the pulse spread.

For a typical NA values of multimode fibers of 0.20 to 0.29, the acceptance angle calculated using (2.4) ranges from 11.5° to 17°. If we take the speed of the ray in optical fiber to be about $2 \cdot 10^8$ m/s [11], the dispersion $t_r$ can be calculated from (2.6). The analog bandwidth of the multimode fiber as a function of the length of the fiber is presented in figure 2.5.

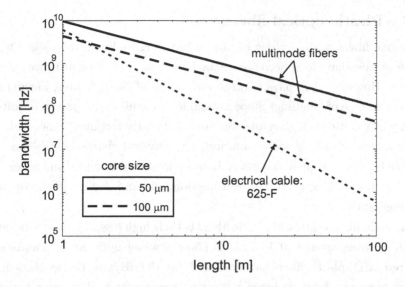

Figure 2.5: The bandwidths of two multimode fibers (core diameters 50 µm and 100 µm) and of an electrical cable as a function of the fiber/cable length.

As far as electrical cables are concerned, the attenuation $A_{tt}$ in dB is proportional to the length of the cable and square-root of the frequency [12, 13]:

$$A_{tt} = e^{-3k_1 l \sqrt{f}} e^{-3k_2 l f} \qquad (2.7)$$

where $f$ is the frequency expressed in megahertz, $k_1$ and $k_2$ are parameters defining the electrical cable type and $l$ is the cable length expressed in kilometers. The first exponential term is due to the skin-effect and the second exponential term is due to the dielectric loss. One should notice that the additional advantage of optical fibers is that the fiber-loss is independent of frequency over their normal operating range [11].

For a very small attenuation cable 625-F [13], $k_1 = 0.6058$ and $k_2 = 0.0016$. Since $k_1 \gg k_2$, the bandwidth of the cable $f_{cab}$ is:

$$f_{cab} = \frac{1}{400 k_1^2 l^2} \qquad (2.8)$$

The behavior of the 625-F cable bandwidth is shown in figure 2.5. For larger transmission distances, the bandwidth of the electrical cable drops significantly in comparison with the bandwidth of the multimode fibers.

### 2.3.3   Plastic optical fibers

Multimode fibers made entirely of plastic have higher losses than silica fibers. Therefore, they have long been outweighed, especially for long distance communication. However, they have also the advantage of being lighter, inexpensive, flexible, and ease of handling. Since the single-mode fibers are proven unsuitable for LAN installations (high connectors cost and costly technical expertise) plastic fibers appear to be a viable solution: the physical characteristics meet the same challenges as copper and glass. It has the ability to withstand a bend radius of 20 mm with no change in transmission, an 1 mm bend without breaking or damaging the fiber.

The main disadvantage of plastic fibers is their high loss. The best laboratory fibers have losses around 40 dB/km. At 650-nm wavelength (for communication using red LED) plastic fibers have loss of about 150 dB/km. Unlike glass-fibers, the loss of plastic fibers is lower at shorter wavelength and is much higher in the near infrared, as illustrated in figure (2.6). As a result, plastic optical fibers have only limited application: they are used mainly for flexible bundles for image transmission and illumination, where light does not need to go far. In communication, plastic fibers are used for short links, like within the office building or cars.

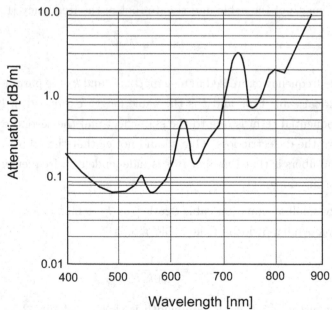

Figure 2.6: Attenuation versus wavelength for a commercial plastic multimode step-index fiber [11]. It typically decreases with wavelength while for the single-mode fibers it increases.

Another important concern is long term degradation at high operation temperatures. Typically, plastic fibers can not be used in applications where the temperature ranges up to 85°C. This leaves only a little margin with engine compartments of car which can get hotter. Plastic fibers are designed similar to glass- fibers;high index cladding (see figures 2.3 and 2.4) encapsulates the low-index core. Commercial plastic fibers are usually multimode.

## 2.4 High intensity light sources

Light source in the fiber-optic communication system converts an electrical input signal into an optical signal. The important parameters of the source are:

- the dimension of the light-emitting area and the radiation pattern of the optical bundle

- the efficiency

- the lifetime

- the effect of temperature on its transfer characteristics

Typical high-intensity light sources are lasers and LEDs. In this work we aim at short distance communications, for which relatively low wavelengths are used: typically around 850 nm[3].

### 2.4.1 Lasers

Vertical cavity laser (VCSEL) are realized by sandwiching a light-emitting semiconductor diode between multi-layer crystalline mirrors. The technologies used for VCSEL fabrication are typically InGaN or AlGaAs. Unlike edge-emitting lasers, which require a larger wafer area and power consumption, the laser output from a VCSEL is emitted from a relatively small area (5-50 $\mu m^2$) on the surface of the chip, directly above the active region. A VCSEL is shown in figure 2.7. The VCSELs physical structure yields numerous inherent advantages including: compact size and surface area, high reliability, flexibility in design, ability to efficiently test each die while still in the wafer state, low current re-

---

[3]Long distance communications uses (expensive) lasers operting at 1300 nm and 1550 nm.

quirements, efficient fiber coupling, high speed modulation, and the ability to build multiple lasers on a single semiconductor. A big advantage of VCSELs is that they can be modulated with very high frequencies (>50 GHz).

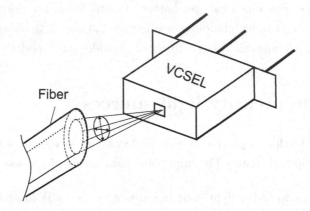

Figure 2.7: VCSEL structure with light emitted from the surface of the chip. Possible coupling with both the single-mode and multimode optical fibers.

### 2.4.2   Light Emitting Diodes (LEDs)

The working principle of the LED is based on emission of photons due to recombination of holes and electrons. The number of carriers present in the active LED region is proportional to the forward current through the LED. The dimensions of the emitting area of an LED are similar to the core diameter of a multimode fiber.

In most LEDs the light is not completely monochromatic i.e. show relatively broad spectra. The visible light from an LED can range from infrared (at a wavelength of approximately 850 nanometers) to blue-violet (about 400 nanometers).

## 2.5   Photodetectors - introduction

A silicon photodetector is in general a solid state transducer used for converting light energy into electrical energy. The following subsections present the main photodetector characteristics.

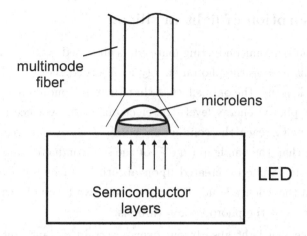

Figure 2.8: A LED coupled to a multimode fiber.

## 2.5.1   Ideal photodetector

In the ideal case, the photodetector should meet the following requirements:

- detect all incident photons,

- has a bandwidth larger than the input signal bandwidth,

- not introduce additional noise, apart from the quantum shot-noise from the received signal.

In most practical applications, additional requirements can be defined. The photodetector should be small, reliable, its characteristics should not be affected by age and environment and it must be cost-effective.

The requirements for ideal photodetectors are very hard to meet in reality, and the photodetectors usually have limited bandwidths with finite response time. They introduce unwanted noise and the efficiency of detecting incident photons is less then 100%. The lifetime is usually limited and some detectors degrade unacceptably as they age.

Most of the photodetectors used in the today's communications are photon-effect based i.e. they directly generate the photocurrent from interactions between the photon and the semiconductor material. Photodetectors are grouped into four categories: photo-multipliers, photoconductors, photodiodes and avalanche photodiodes. In this book the main focus will be on photodiodes. The limitations of photodiodes in standard CMOS in their quantum efficiency and in the bandwidth will be discussed in the following chapters.

## 2.5.2   Absorption of light in silicon

Light shining onto a semiconducting material is absorbed in that material. More precisely, in this process the photon energy is absorbed. For low photon energy (i.e. long wavelengths) the only effect is that the semiconductor material heats up. For higher photon energy levels the electrons in the valence band may get sufficient energy to reach the conduction band. Clearly this requires photon energies larger than the bandgap (in eV) of the semiconductor material. In this last case, the single photon created upon absorbtion a mobile electron and a mobile holes in the valence band. Basically, these two types of carriers are seen as a photocurrent at the photodiode terminals.

In the process of light absorbtion, over a certain distance into a material a (material and wavelength related) fraction of the photons is absorbed. The result is then that the light-intensity decreases exponentially with distance into the material [8]. In equation:

$$I \propto e^{-\alpha x} \tag{2.9}$$

where $\alpha$ is the wavelength (and material) dependent absorption coefficient while $x$ is the depth in silicon. The absorption coefficient for silicon can be approximated with the following formula [14]:

$$\alpha = 10^{13.2131 - 36.7985\lambda + 48.1893\lambda^2 - 22.5562\lambda^3} \quad 1/[\text{cm}] \tag{2.10}$$

The wavelength $\lambda$ of the input light signal is given in [µm].

Photodiodes in CMOS technology are sensitive only for a particular wavelength range. The photon energy $h\nu$ is wavelength dependent and it should be larger than the bandgap of the semiconductor material (in this case silicon) [15]. For relatively large wavelengths the photon energy is not high enough to create an electron-hole pair in silicon; for silicon this is for $\lambda > 950$ nm. For lower wavelengths on the other hand, $\lambda < 400$ nm, excess carriers are generated very close to the photodiode surface. Because typically the surface recombination rate is high then only a small part of the generated carriers contribute to the photocurrent, the usable wavelength sensitivity range of CMOS photodiodes is $\lambda \in [400 - 850]$ nm.

For best performance e.g. the highest speed and responsivity, the photodiode should be designed to allow the largest number of photons to be absorbed in depletion regions; in the ideal case photons should not be absorbed until they have penetrated as far as the depletion region, and should be absorbed before penetration beyond it. The relative depth to which photon penetrates is a function of its wavelength (see chapters 3, 4 and 5). Short wavelength light (around blue and violet) are absorbed close to the photodiode surface while those with longer wavelength (infrared) may penetrate 10ths of micrometers deep in the substrate.

The values of the absorption coefficient and the corresponding $1/e$-absorption depths[4] in silicon, are shown in figure 2.9. From this figure we conclude that the difference in absorption coefficient for the two boundaries is very large: $\alpha = 7.5 \times 10^2 \div 5.5 \times 10^4 \quad \text{cm}^{-1}$. As a result, the difference in $1/e$-absorption depths for 400 nm and 850 nm light is almost three orders of magnitude.

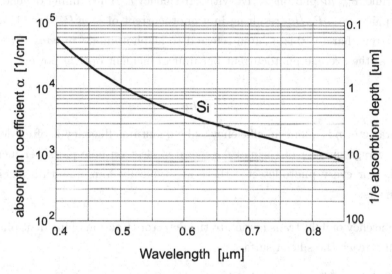

Figure 2.9: The absorption coefficient $\alpha$ for silicon photodiodes versus input wavelength of the light signal $\lambda$.

The light intensity drops exponentially inside silicon:

$$\frac{\partial I}{\partial x} \propto \alpha e^{-\alpha x} \tag{2.11}$$

The more light is absorbed in the photodiode, the more excess carriers are generated. We define a parameter $G(x)$ which is the *carrier generation rate* as

---

[4]The $1/e$ absorption depth is the depth into the silicon for which the light-intensity is dropped to $1/e$ of the incidentlight-intensity. This depth is equal to $1/\alpha$ of the input wavelength

a result of the incident light in the unity of time often modelled as:

$$G(x) = \Phi_0 \alpha e^{-\alpha x} \tag{2.12}$$

where $\Phi_0$ is the *photon flux* at the silicon surface generated by a monochromatic optical source and can be further expressed as:

$$\Phi_o = \frac{P_{\text{in}}}{h\nu}(1 - R_{\text{f}}) \tag{2.13}$$

$P_{\text{in}}$ is the input optical power density (W/cm$^2$), $h\nu$ is the photon energy and $R_{\text{f}}$ is the reflection coefficient due to the different index of reflections of the "outside world" on the top of the silicon and the silicon itself [8]. During each unit of time, $P_{\text{in}}/h\nu$ photons arrive with a frequency $\nu$. The number of generated carrier pairs is $\sim \eta P_{\text{in}}/h\nu$ resulting in a photocurrent of $\sim \eta e P_{\text{in}}/h\nu$ [8] (where $e$ is electron charge); this is often referred to as photodiode *responsivity*. It is defined as the average photocurrent per unit of incident optical power:

$$R = \frac{e\eta}{h\nu} \tag{2.14}$$

The parameter $\eta$ is quantum efficiency. The quantum efficiency is often defined as the average number of (primary) generated electron-hole pairs per incident photon. For every photodetector there are typically four quantum efficiency components:

1. efficiency of light transmission to the detector (fraction of incident photons that reach the silicon surface)

2. efficiency of light absorption by the detector (fraction of photons reaching the silicon surface that produce electron-hole (EH) pairs)

3. quantum yield (number of EH pairs produced by each absorbed photon)

4. charge collection efficiency of the photo-detector (fraction of generated minority carriers by presence of light, that cross the pn junction before recombining).

However, during the calculations of the available output photocurrent, typically only the first and the fourth quantum efficiency components are taken into account. The other two components are taken to be equal to one. Typical value of the quantum efficiency in a CMOS photodiode is about 40%-70%.

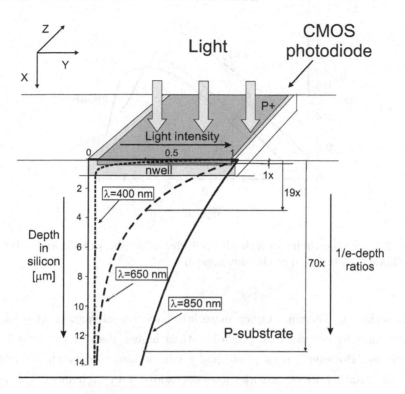

Figure 2.10: The absorption of light inside photodiode in standard CMOS technology. The difference between 1/e-absorption depth among $\lambda = 400$, 650 and 850 nm) is large; There is a causal relation between the photodiode responsivity and the bandwidth.

The maximum possible responsivity varies with photon energy. For $\eta = 1$, the maximal responsivity can be simplified as: $R_{max} = \lambda/1.24$, where $\lambda$ in [µm]. For the wavelength sensitivity range of CMOS photodiodes 400 nm$<\lambda<$850 nm, the maximum responsivity is in the range 0.32 A/W$<R_{max}<$0.64 A/W.

The responsivities of a typical Si photodiode, Ge photodiode and InGaAsP photodiode as a function of wavelengths are shown in figure 2.11. In that figure, the maximum responsivity is marked by the line indicated with $\eta = 1$.

In the short-wavelength region ($\lambda = 400$ nm), the value of $R_{max}$ decreases more rapidly than $\lambda$; this is caused by increased surface recombination for the shallow absorption depth. For large wavelengths ($\lambda>850$ nm) the responsivity of the CMOS photodiodes also declines; minority carriers are generated deep in the substrate and they are recombined with majority carriers.

Figure 2.11 shows that silicon photodiodes are not useful in the longer wave-

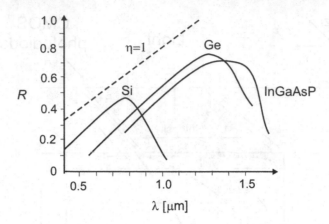

Figure 2.11: Responsivity of a Si photodiode, a Ge photodiode and a InGaAs photodiode as a function of the wavelength

length region $\lambda$>950 nm.  Other materials have the advantage of a smaller bandgap and higher mobility providing thus higher responsivity and higher bandwidths. However, silicon photodiodes can be integrated with mainstream electronic circuitry which provides low-cost solution for high-speed optical detection.  This last point is the main motivation for the work presented in this book.

## 2.6   High-speed optical receivers in CMOS for $\lambda = 850$ nm-literature overview

This section presents a brief overview of high-speed optical receivers in CMOS technology reported in the literature for $\lambda = 850$ nm.  Only a few solutions for optical receivers are reported in standard CMOS; the reported data-rates in standard CMOS is up to 700 Mb/s. Other publications use modified CMOS technology and high-voltage solutions with reported data-rates up to 1 Gb/s.

### 2.6.1   Using standard CMOS technology

High-speed optical detection is typically achieved in two manners.  Firstly "smart" photodiode and full exploitation of the possibilities in a technology can be done. These possibilities include layout issues, using high voltages, adding processing features and more. Secondly, slow standard photodiodes can be used, with electronic postprocessing to boost speed.

### CMOS technology with feature size of 1μm

In [16], a data-rate of 622 Mb/s is achieved in a 1-μm CMOS technology with a diode bias voltage of 5 V and with 850 nm light. The reported sensitivity of the detector is -15.3 dBm for a bit error rate (BER) of $10^{-9}$ which is low compared to the requirements for e.g. the Gigabit Ethernet Standard: -17 dBm for the same BER [17].

Important differences between a typical 1 μm CMOS processes and a 0.18 μm CMOS process (used as demonstrator process in this book) include:

- the depth of the nwell is about 4 μm which is 3-4 times larger than in modern CMOS technology.

  For $\lambda = 850$ nm, a large portion of light (roughly 1/3) is then absorbed in nwells, in comparison with newer CMOS technologies where over 80% of the light is absorbed inside the substrate. As a direct result, the (fast) diffusion inside the nwell contributes significantly to the speed of the photodiode in the $1\mu m$ process; in modern CMOS typically the (slow) bulk currents are far dominant. A full analysis of speed aspects is given in chapter 3

- the supply voltage is almost three times higher (5V/1.8V); as a result the depletion region width is about 50% higher which again gives the photodiode in a 1 $\mu m$ process a speed advantage over diodes in 0.18 $\mu m$ processes.

- 1 μm CMOS is outdated, and cannot implement electronic circuits in the GHz range.

Together with the depletion region that has a couple of μm depth inside the epi-layer, the amount of the carriers that are generated deep in the substrate is 5 times lower than in modern CMOS technologies[5]. For comparison, the photodiode bandwidth for a modern CMOS process (0.18 μm) is only 1 MHz for $\lambda = 850$ nm (see chapter 3).

---

[5]Slow diffusion of the substrate carriers that limit the photodiode bandwidth is tremendously reduced (exponential light absorbtion). This will be discussed in detail in chapter 3.

## SML detector exploiting layout design

One solution in standard 0.25 μm CMOS technology where 700 Mb/s data-rate is achieved is presented in [18, 19]. The effect of the slowly diffusing carriers is cancelled by subtracting two diode responses: one immediate and one deferred diode responses.

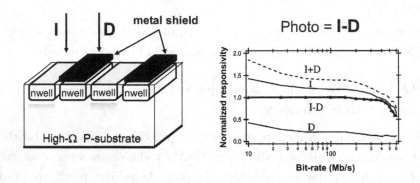

Figure 2.12: Spatially modulated light detector.

The principle of the SML-detector allows one to cancel the effect of the substrate carriers at the cost of lower responsivity. The SML-detector consists of a row of rectangular p-n junctions (fingers) alternatingly covered and non-covered with a light blocking material, as shown in figure 2.12. The masked fingers connected together form the *deferred* (D) detector. The other fingers connected together form the *immediate* (I) detector.

The slow tail in the time-response of both detectors is very similar, since approximately the same number of the substrate carriers diffuse towards the two detectors. The fast overall photodiode response is achieved by subtraction of the two diode responses. This however results in lower responsivity (about 75% of the input signal is lost) and hence lower sensitivity. For 300 Mb/s data-rate and BER=$10^{-9}$ the reported sensitivity was -18 dBm. The detector responsivity for 700 Mb/s [19] was not reported; typically the optical power of the input signal is even higher since the noise in the circuit is increased for higher speeds.

## 2.6.2  CMOS technology modification

### Very high-resistance substrate

A solution for 1 Gb/s optical detection is presented in [20]. An integrated receiver is designed in NMOS technology with a special high-resistive substrate which behaves as a diode intrinsic (I) region. This PIN photodiode is used as a detector designed using n+ and p+ layers inside high-resistive n-substrate. A large intrinsic region ensures both the high speed and the high quantum efficiency of 82%. However, the supply voltage is -32 V. This is unrealistic biasing in modern CMOS processes where typical supply voltage is around 1 V.

### Buried oxide layer

In order to increase the photodiode bandwidth, the dominant slow substrate diffusion current [18] can be cancelled by introducing an *buried oxide layer*. The working principle is similar with silicon-on-isolator (SOI) photodetectors. The biggest disadvantage of this technique is a reduced responsivity. The large portion of the excess carriers generated in the substrate do not contribute to the overall photocurrent. In [21], a bandwidth of 1 GHz is reported with the cost of very low[6] responsivity of 0.04-0.09 A/W, corresponding to a sensitivity of 2 dBm to -5 dBm. As a result, the input optical power should be at least 13 dB higher than required in Gigabit Ethernet Standard [17].

---

[6]Typically, responsivity of the photodiode is $> 0.3$ A/W corresponding to $>40\%$ quantum efficiency.

# Bibliography

[1] G. P. Agrawal, N.K. Dutta: *"Long-Wavelength Semiconductor Lasers"*, Van Nostrand Reinhold, New York, 1986.

[2] G. P. Agrawal: *"Fiber-Optic Communication Systems"*, John Wiley and Sons, New York, 1997.

[3] R. Goff: *"Fiber Optics Reference Guide"*, Focal Press, Third Edition, 2002.

[4] D. A. B. Miller: "Physical reasons for Optical interconnection", *Int. J. Optoelectronics* 11, pp. 155-168, 1997.

[5] D. A. B. Miller and H. M. Ozaktas:" Limit to the bit-rate capacity of electrical interconnects from the aspect ratio of the system architecture", *Journal of parallel Distrib. Comput.*, vol 41, p4242, 1997.

[6] D. A. B. Miller: "Rationale and challanges for optical interconnects to electronic chips", *Proc. IEEE*, vol. 88, 2000, pp.728-749.

[7] H. B. Bakoglu: *"Circuits, interconnections, and packaging for VLSI"*, Addison-Wesley, New York, 1990.

[8] W. Etten and J. vd Plaats: *"Fundamentals of Optical Fiber Communications"*, Prentice-Hall, 1991.

[9] M. R. Feldman, S. C. Esener, C. C. Guest and S. H. Lee: "Comparison between electrical and optical interconnect based on power and speed consideration", *Appl. Optics*, vol. 27, pp. 1742-1751, 1998.

[10] E. A. de Souza, M. C. Nuss, W. H. Knox and D. A. Miller: "Wavelength-Division Multiplexing with femtosecond pulses", *Optics Letters*, vol. 20, pp. 1166-1168, 1996.

[11] J. Hecht: "*Understanding Fiber Optics*", Prentice Hall, New Jersey, 1999.

[12] A. Bakker: "An Adaptive Cable Equalizer for Serial Digital Video Rates to 400 Mb/s", *Dig. Tech.Papers ISSCC 1996*, pp. 174-175.

[13] W. Chen: "*Home Networking Basis: Transmission Environments and Wired/Wireless Protocols*", Prentice Hall PTR., July 2003.

[14] W. J. Liu, O. T.-C. Chen, L.-K. Dai and Far-Wen Jih Chung Cheng: "A CMOS Photodiode Model", *2001 IEEE International Workshop on Behavioral Modeling and Simulation*, Santa Rosa, California, October 10-12, 2001.

[15] S. M. Sze: "*Physics of semiconductor devices*", New York: Wiley Interscience, 2-nd edition, p. 81, 1981.

[16] H. Zimmermann and T. Heide: "A monolithically integrated 1-Gb/s optical receiver in 1-μm CMOS technology", *Photonics technology letters*, vol. 13, July 2001, pp. 711-713.

[17] IEEE 10 Gigabit Ethernet Standard 802.3ae.

[18] D. Coppée, H. J. Stiens, R. A. Vounckx, M. Kuijk: "Calculation of the current response of the spatially modulated light CMOS detectors", *IEEE Transaction Electron Devices*, vol. 48, No. 9, 2001, pp. 1892-1902.

[19] C. Rooman, M. Kuijk, R. Windisch, R. Vounckx, G. Borghs, A. Plichta, M. Brinkmann, K. Gerstner, R. Strack, P. Van Daele, W. Woittiez, R. Baets, P. Heremans: "Inter-chip optical interconnects using imaging fiber bundles and integrated CMOS detectors", *ECOC'01*, pp. 296-297.

[20] C. L. Schow, J. D. Schaub, R. Li, and J. C. Campbell: "A 1 Gbit/s monolithically integrated silicon nmos optical receiver", *IEEE Journal Selected Topics in Quantum Electron.*, vol. 4, Nov.Dec. 1999, pp. 1035 1039.

[21] M. Ghioni, F. Zappa, V. P. Kesan, and J.Warnock: "VLSI-compatible high speed silicon photodetector for optical datalink applications", *IEEE Trans. Electron. Devices*, vol. 43, July 1996, pp. 1054 1060.

# CMOS photodiodes for $\lambda = 850$ nm

*This chapter presents frequency and time domain analyses of photodiode struc-
tures designed in standard CMOS technology, for $\lambda = 850$ nm. For clear ex-
planation and illustration of the physical processes inside photodiodes, one particular CMOS technology is analyzed in detail: a standard 0.18 μm CMOS[1].
The photodiodes are first analyzed as stand-alone detectors. This allows the
analysis of the intrinsic photodiode behavior, related to the movement (drift and
diffusion) of the generated carriers inside the diode. In the second part of this
chapter, the diode is investigated as an "in-circuit" element, integrated together
with the subsequent electronics. The electrical bandwidth of the photodiode is
determined by the diode capacitance and the input impedance of the subsequent
amplifier. These two bandwidths determine the total diode bandwidth. Further,
the influence of the diode layout (nwell, n+, p+ finger sizes) in general, on the
intrinsic, the extrinsic and the total bandwidth is investigated.*

---

[1]Choosing another CMOS technology does not fundamentaly change the behaviour of the
photodiode in general. The impact of the technology on photodiode behavior is discussed in
detail in chapter 6.

# 3.1   Introduction

Depending on the wavelength of the input optical signal (400 nm$\leq\lambda\leq$850 nm), there are several applications for optical detectors:

- $\lambda = 850$ nm: 10 Gigabit/sec Fiber Ethernet (standard 802.3ae, [1]), short-haul communication (chip-to-chip, board-to-board), high-speed opto-couplers [2].

- $\lambda = 780$ nm: CD players and recorders

- $\lambda = 650$ nm: DVD players and recorders

- $\lambda = 400$ nm: DVD - blue ray disc

- 400 nm$\leq\lambda\leq$700 nm: CMOS image sensors

This chapter shows that the bandwidth of the integrated CMOS photodetectors is wavelength dependent, structure dependent and layout dependent. It is important to notice that the technology used in book is standard CMOS; there are no technology modifications. *The depth* of the photodiode regions where light is absorbed is related to the photodiode structure; in this chapter various photodiode structures are studied in detail:

- nwell/p-substrate

- n+/p-substrate

- p+/nwell/p-substrate

- p+/nwell

The size of the lateral depletion region in photodiodes depends on the well-technology. Therefore, the total depletion region contribution to the overall photocurrent is also different. Two well technologies are analyzed in this chapter:

- *twin-well with adjoined wells* and

- *triple-well with separate wells* technology.

For $\lambda$=600-850 nm, the light penetration depth is larger than 6 $\mu$m (90% of the absorbed light, [2][2]. Only 10% of the light is absorbed in the wells and junctions

---

[2]For modern CMOS technologies, for example 0.18 $\mu$m CMOS, the deepest junctions are located close to the photodiode surface (typically 1-2 $\mu$m)

while 90% is absorbed in the substrate. Two different kinds of p-substrate are analyzed:

- high-resistance substrate

- low-resistance substrate

*The width* of the photodiode regions located close to the surface (nwell, n+, depletion regions) can be optimized for the best photodiode performance (maximal bandwidth and responsivity). The diode can comprise a number of nwells, n+ fingers, or it can be designed as a single photodiode i.e. with maximal nwell/n+ width. This is illustrated in figure 3.1. The influence of the nwell/n+/p+ geom-

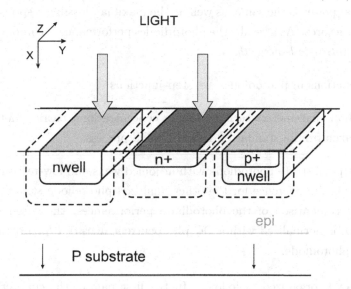

Figure 3.1: Photodiode structures in standard CMOS technology.

etry on photodiode bandwidth will be derived. Two different geometries will be discussed in detail:

- minimal nwell/n+ width $L_{ymin}$. Typically for standard CMOS, the minimal width is twice the nwell/n+ depth; for 0.18 µm CMOS, $L_{ymin}=2$ µm.

- nwell/n+ width much larger than the nwell/n+ depth: $L_y=10$ µm.

In these sections $L_y$ is the width of the nwell/n+/p+ regions. The comparison of different photodiodes in this book is based on the intrinsic $FOM_i$ measure introduced in section 3.3. This FOM$_i$ has analogies with the well known gain-bandwidth product, but takes roll-off properties of photodiode into account. For fair comparison of various photodiodes, the following assumptions are taken:

- fingered photodiodes are considered in general, to have nwell/n+ stripes with different sizes. The nwell/n+ regions are placed at *minimal distance* defined by the technology[3]. In this manner, the junction area is maximized for fixed width stripes.

- the active (light sensitive) area of all diodes is identical. The active area corresponds to the illuminated silicon area. Therefore, the absorbed input optical power is the same, as well as the maximal possible responsivity for all structures. As a result, the photodiodes performance is compared using their *intrinsic bandwidth*.

- all junctions in photodiodes are step-junctions

- the diode parameters such as doping concentrations, carrier lifetime etc. are taken from a standard 0.18 µm CMOS process.

- nwell/p-substrate photodiode with adjoined wells and low-resistance substrate is the reference for the other analyzed photodiode structures. For easier comparison of the photodiode performances, their frequency response is normalized with a DC photocurrent density ($J_{DC}$) of the reference photodiode.

This chapter is organized as follows. In the first part of the chapter, a photodiode is analyzed as a stand-alone device. The intrinsic (physical) behavior of CMOS photodiodes for the aforementioned diode structures and layouts is analyzed. The analyses use the calculated frequency and time responses of the diode with Dirac pulse as the input optical signal. Calculations of the drift current profile in the depletion region and the diffusion current profiles in the remaining n- and p-regions are presented. The overall photocurrent is the sum of the drift and the diffusion currents. The frequency behavior of the overall photocurrent gives insight into the maximum intrinsic bandwidth limitations of various photodiodes. The photodiodes intrinsic figure-of-merit FOM$_i$ will be introduced.

---

[3]Larger distances between nwells decrease a carrier-gradient of the excess carriers inside p-regions i.e. decrease the diffusion speed of these carriers and limits the total diode bandwidth.

In the second part of the chapter, the photodiode is investigated as an "in-circuit" element, integrated together with the subsequent transimpedance amplifier (TIA). The diode capacitance and TIA's input capacitance together with the TIA's input resistance gives an extrinsic bandwidth. This bandwidth will be here referred to as *electrical* diode bandwidth. For a constant input resistance of the TIA, the larger the diode capacitance the smaller the electrical bandwidth. This capacitance is directly related to the diode layout i.e. related to the nwell size. This chapter shows the trade-off in the diode layout design for the maximal *total* photodiode bandwidth. The photodiodes extrinsic figure-of-merit $\text{FOM}_{\text{ex}}$ will be introduced.

# 3.2   Bandwidth of photodiodes in CMOS

The main topic of the chapter is the frequency response of various photodiodes designed in 0.18 µm CMOS [6]. Using a novel figure-of-merit for photodiode behaviour, the most suitable photodiode can be chosen for a certain application prior to the circuit fabrication and testing.

First, a nwell/p-substrate diode with high-resistance substrate and separate-well technology is analyzed both in the frequency domain and in the time domain. The high-resistivity substrate is chosen for simplicity of calculation. After this, the frequency and time responses of nwell/p-substrate photodiode with *low-resistance* substrate will be derived. This low-resistance substrate is used for photodiode fabrication, and for this reason the latter diode structure will serve as the reference for the other analyzed photodiode structures.

## 3.2.1   Intrinsic (physical) bandwidth

The intrinsic diode characteristic is related to the behavior of the optically generated excess carriers inside the photodiode. These carriers are moving inside the photodiode either by drift (inside depletion regions) or by diffusion (outside depletion regions). In general, the photodiode response is the sum of the drift current $I_{\mathrm{drift}}$, and the diffusion currents $I_{\mathrm{diff_k}}$:

$$I_{\mathrm{int_{total}}} = I_{\mathrm{diff_{nwell}}} + I_{\mathrm{diff_{n+}}} + I_{\mathrm{diff_{p+}}} + I_{\mathrm{diff_{p-subs}}} + I_{\mathrm{drift}} \qquad (3.1)$$

For better understanding of the total diode response, the frequency response of every current component will be separately presented. The excess carrier profiles and the currents of the different photodiode regions are calculated by taking the Laplace transform of the diffusion equations in the time domain, [2]. These analyses are used to estimate the frequency domain behavior of CMOS photodiodes.

### Nwell/p-substrate photodiode with high-resistance substrate in twin-well technology

This section presents a frequency analysis of the finger nwell/p-substrate photodiode, shown in figure 3.2. The number of fingers, $N$, is determined by the photodiode dimension in the y-direction $Y$ (see figure 3.2), and by the technology (the minimal nwell/pwell finger width $L_{\min}$). This number can take every value in the range:

$$N = \left\lfloor \frac{Y}{L} \right\rfloor \quad \text{where} \quad L \in [L_{\min}, Y] \tag{3.2}$$

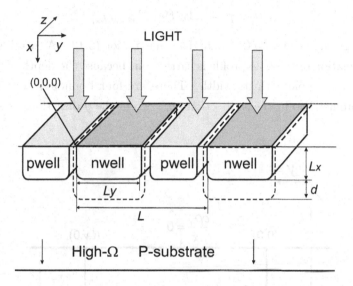

Figure 3.2: Finger nwell/p photodiode structure with *high resistance* substrate and twin-wells in standard CMOS technology.

**Nwell diffusion current**

The nwell diffusion current in (3.1), is solved analytically for an impulse light radiation, in two-dimensions using a method similar to that in [2]. From the hole carrier profile, the current density is calculated at the border of the depletion region since the excess holes are collected there as a photocurrent.

The transport of the diffusive holes inside the photodiode is described by the diffusion equation [3]:

$$\frac{\partial p_n(t,x,y)}{\partial t} = D_p \frac{\partial^2 p_n(t,x,y)}{\partial x^2} + D_p \frac{\partial^2 p_n(t,x,y)}{\partial y^2} - \frac{p_n(t,x,y)}{\tau_p} + G(t,x,y) \tag{3.3}$$

where $p_n(t, x, y)$ is the excess minority carrier concentration inside the nwell, $D_p$ is the diffusion coefficient of the holes in the n-doped layer and $\tau_p$ is the minority-carrier lifetime. Using (2.12) the hole generation rate $G(t, x, y)$ can be expressed as:

$$G(t, x, y) = \alpha \Phi_0(t) e^{-\alpha x}\big|_{x \in [0, L_x]} \tag{3.4}$$

where $\alpha$ is given in equation (2.10), and $\Phi_0$ in equation (2.13). A two-dimensional $(x, y)$ calculation of the hole-profile is carried out because the depth of the nwell region is comparable with its width. There are four boundary conditions for the hole-profile: two in the $x$-direction and two in the $y$-direction, as shown in figure 3.3.

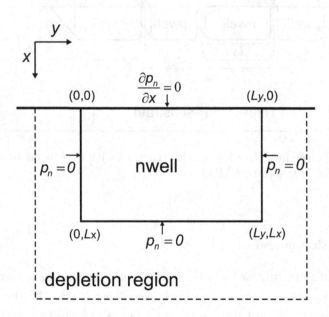

Figure 3.3: The boundary conditions for hole densities inside nwell.

For the *first* boundary condition, the photodiode surface is assumed to be reflective i.e. the normal component of the gradient of the carrier density is zero. This is because the surface recombination process is slow compared to the timescale used for Mb/s and Gb/s datarates as considered in this book. On the *other three* boundaries with depletion region, the electron densities are assumed to be zero:

$$\frac{\partial p_n}{\partial x}\Big|_{x=0} = 0 \quad p_n|_{x=L_x} = 0 \tag{3.5}$$

$$p_n|_{y=0} = 0 \quad p_n|_{y=L_y} = 0 \tag{3.6}$$

Equation (3.3) is a partial differential equation in time ($t$) and space domain ($x, y$). The hole profile $p_p$ is calculated first by taking the Laplace transform of the diffusion equation (3.3). In this manner the carrier profile is transformed to the frequency domain. The obtained diffusion equation is solved in the space domain ($x, y$). In order to solve this equation analytically, the most suitable method is to use Discrete Fourier series in the space domain. This is certainly valid for $y$-direction where nwell/pwell represents light/no-light periodic function. The carrier distribution function $p_p$ and the carrier generation function $G$ are rewritten as a product of two Fourier series; one of a square wave in the $x$-direction (with index $n$) and the other of a square wave in the $y$-direction (with index $m$). Each of these decomposed terms of $G(s)$ drive one of the terms of decomposed of $p_n$:

$$\frac{p_n(s, x, y)}{\Phi_0(s)} = \frac{16\alpha L_y L_p^2}{l\pi D_p} \sum_{n=1}^{\infty} \sum_{m=1}^{\infty} \frac{(2n-1)\pi(-1)^{\frac{n-1}{2}}e^{-\alpha L_x} + 2\alpha L_x}{4\alpha^2 L_x^2 + (2n-1)^2\pi^2}$$

$$\times \frac{\sin(\frac{(2m-1)\pi y}{L_y})\cos(\frac{(2n-1)\pi x}{2L_x})}{(2m-1)\left(\frac{(2n-1)^2\pi^2 L_p^2}{L_y^2} + \frac{(2m-1)^2\pi^2 L_p^2}{4L_x^2} + s + 1\right)} \tag{3.7}$$

The Fourier series is composed of odd sine and cosine terms. The even Fourier terms (integer number of sines and cosines) do not contribute to the nwell current response because the total area below the curves is zero. The odd sine and cosine terms are truncated, and the area below these curves is non-zero, see figure 3.4.

Once the carrier profile is calculated, the hole-current frequency response can be determined for each set of indexes $n$ and $m$ (from equation (3.7)). The total contributed current is the integral of the current through the two side-walls and the bottom layers. The final expression for the nwell diffusion current is:

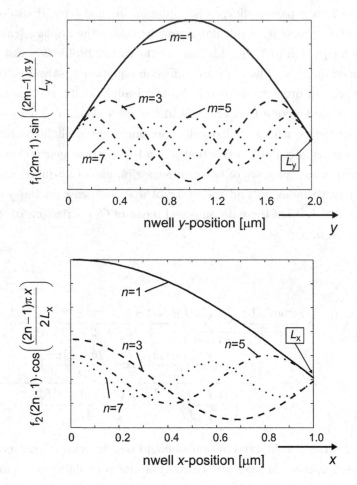

Figure 3.4: Minority carrier profile inside nwell in a) $x$-direction and b) $y$ directions.

$$\frac{J_{\text{nwell}}(s)}{\Phi_0(s)} = 32\frac{eL_{\text{p}}^2\alpha}{l\pi^2}\sum_{n=1}^{\infty}\sum_{m=1}^{\infty}\frac{(2n-1)\pi e^{-\alpha L_x}+(-1)^{\frac{(2n-1)-1}{2}}\alpha L_x}{4\alpha^2 L_x^2+(2n-1)^2\pi^2}$$

$$\times \ \frac{\dfrac{2L_x}{L_y}\dfrac{1}{2n-1}+\dfrac{L_y}{2L_x}\dfrac{2n-1}{(2m-1)^2}}{\dfrac{(2n-1)^2\pi^2 L_{\text{p}}^2}{4L_x^2}+\dfrac{(2m-1)^2\pi^2 L_{\text{p}}^2}{L_y^2}+1+s\tau_{\text{p}}} \tag{3.8}$$

The total nwell response is the double sum of the $n$ and $m$ one-pole responses with wavelength-dependent amplitudes. The nwell amplitude response is shown in figure 3.5 and the nwell phase response in figure 3.6. The total response is shown in figure 3.7. The former figure shows that the amplitude of higher Fourier terms decreases with $n$ and $m$ while the poles are placed further on the frequency axes. The sum of all components gives the total nwell response with an unusually *low roll-of* ($\sim$10 dB/decade). This is a feature of the nwell diffusion process that will be taken advantage of in chapter 4.

The bandwidth of the nwell diffusion current can be estimation from (3.8) using certain simplifications given in [2]. The slowest and the dominant contribution to the nwell current corresponds to the case $n = m = 1$. The amplitude of the other contributions decrease quadratically with $n$ and $m$. From (3.8), the -3 dB bandwidth frequency can be approximated with:

$$f_{\text{3dB}} \simeq \left(\frac{\lambda}{\lambda_{850}}\right)^{1/3}\frac{\pi D_{\text{p}}}{2}\left(\left(\frac{1}{2L_x}\right)^2+\left(\frac{1}{L_y}\right)^2+\left(\frac{1}{L_{\text{p}}}\right)^2\right) \tag{3.9}$$

The only difference in the equation above in comparison with the bandwidth equation given in [2] is that wavelength ($\lambda$) dependence is introduced. Equation (3.9) shows that the bandwidth of the nwell current is directly proportional to the diffusion constant of holes $D_{\text{p}}$ and therefore to the mobility of holes [3]. The higher the mobility, the faster the holes reach the edges of the depletion region and the faster the response.

The terms between brackets in (3.9) concerning the depth $L_x$ and the width $L_y$ can be explained using figure 3.8 and figure 3.9. The latter presents the time response of the nwell region calculated by using the inverse Laplace transform of equation (3.8). The holes diffuse towards junctions due to the gradient of the hole concentration. The gradient is maximum in the direction of the minimum

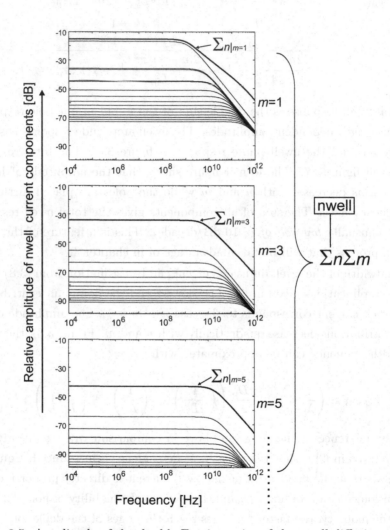

Figure 3.5: Amplitude of the double Fourier series of the nwell diffusion current for the vertical and the lateral nwell direction. The total nwell diffusion current is the sum of all terms with indices $m$ and $n$.

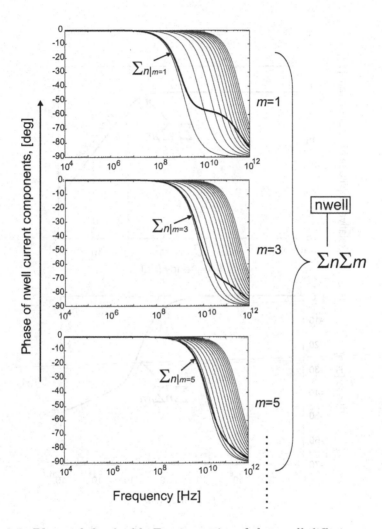

Figure 3.6: Phase of the double Fourier series of the nwell diffusion current for the vertical and the lateral nwell direction. The total nwell diffusion current is the sum of all terms with indices $m$ and $n$.

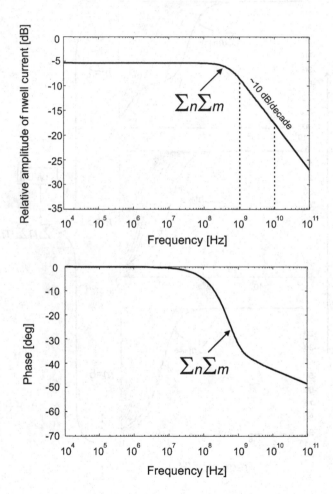

Figure 3.7: The total amplitude and phase of the nwell diffusion current.

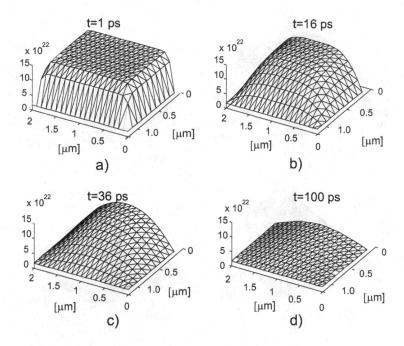

Figure 3.8: The calculated hole diffusion profile inside nwell with 2μm size, under incident light pulse (10 ps pulse-width). This profile is calculated after 1 ps, 16 ps, 36 ps, 100 ps.

distance to the junctions. Therefore, the holes tend to choose "minimal paths" towards the junctions. If $L_x=2L_y$ the hole in the top-middle position of the nwell can diffuse left, right or down with the equal probability since they are all "minimal paths". The nwell size in $y$-direction is twice the size in $x$-direction and for that reason the first bracket term is with $1/(2L_x)$.

The third term inside the brackets corresponds to the diffusion length of the holes $L_p$. Typically in CMOS technology, the diffusion length is much larger than the minimum side of the nwell and its contribution in the equation (3.9) is small. It is clear that both the *layout* (lateral size) and the *technology* (related to the nwell depth and the doping concentration) are very important and determine the nwell diffusion bandwidth.

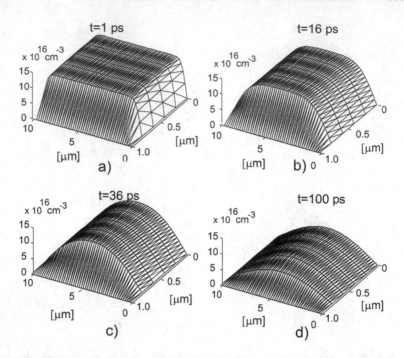

Figure 3.9: The hole diffusion profile inside nwell with 10 µm size, under incident light pulse (10 ps pulse-width). The lateral nwell dimension is obviously less important for the diffusion process.

### High-resistance substrate current

The second photocurrent component analyzed in this chapter is the substrate current. The substrate current is the photocurrent resulting from generated charge below wells and between wells. The diffusion process of electrons generated in the substrate below the depletion regions is different from the diffusion of electrons generated between wells.

Taking into account the depth of wells and the penetration depth of light in silicon, it follows that typically the contributions of generated charge below wells is dominant. Therefore, for simplicity reasons sidewall effects of wells are neglected which effectively approximates a photodiode as a single well device. This simplification yields much simpler derivations at the cost of only a small error.

Generated carriers in the substrate diffuse either towards upper allocated junctions (nwell or n+) or deeper into the substrate where they are recombined. The substrate current component consists of the non-recombined carriers, diffusion upwards. The substrate current frequency response on a Dirac light pulse can be calculated using a one-dimensional (vertical) diffusion equation [3]:

$$\frac{\partial n_p(t,x)}{\partial t} = D_n \frac{\partial^2 n_p(t,x)}{\partial x^2} + \frac{n_p(t,x)}{\tau_n} + G(t,x) \tag{3.10}$$

where $n_p(x,t)$ is the excess electron concentration inside the substrate, $D_n$ is the diffusion coefficient of the electrons and $\tau_n$ is the minority-carrier lifetime. The electron generation rate $G(t,x)$ using equation (2.12) is:

$$G(t,x) = \alpha \Phi_0(t) e^{-\alpha(L_x+d)} e^{-\alpha x}\big|_{x\in[0,L_{\text{fnt}}]} \tag{3.11}$$

where $L_{\text{fnt}}$ is the depth where the light is almost completely absorbed (99%), and $d$ is the depletion region depth (see figure 3.2). To simplify the calculation at this $f_{fnt}$ the excess carrier concentration is approximated to be zero. This simplifies derivations at the cost of only a small error.

There are two boundary conditions for the minority electrons in the $x$-direction; the first boundary is at the substrate top and the second at $L_{\text{fnt}}$. Both boundary conditions are taken to be zero since the carriers are either removed by the junctions or they are recombined:

$$n_p\big|_{x=0} = 0 \tag{3.12}$$

$$n_p\big|_{x=L_{\text{fnt}}} = 0 \tag{3.13}$$

For $\lambda = 850$ nm the chosen bottom boundary $L_{\text{fnt}} = 60$ µm. Larger values for $L_{\text{fnt}}$ leads to a slower response [18].

To solve equation (3.10), the Laplace transform of the equation is taken first, similar to the procedure in [2]. The carrier profile $n_p(t,x)$ is transformed to $n_p(s,x)$ in the frequency domain. The carrier profile function $n_p$ and the carrier generation function $G(s)$ are rewritten as the product of a Fourier series of a square wave in the $x$-direction (with index $n$). Each of these decomposed terms of $G(s)$ corresponds to one of the terms of $n_p(s,x)$. For each set of indexes $n$ a carrier profile is calculated and expressed as:

$$n_p(s,x) = 2\alpha\Phi(s)\pi \sum_{n=1}^{\infty} \frac{n\sin\left(\frac{n\pi x}{L_{\text{fnt}}}\right)}{(\alpha^2 L_{\text{fnt}}^2 + n^2\pi^2)D_n\left(\frac{s}{D_n} + \frac{1}{L_n^2} + \frac{m^2\pi^2}{L_{\text{fnt}}^2}\right)} \tag{3.14}$$

The carrier profile is the sum of $n$-sine signals that differs in amplitude by a factor $n^2/(\alpha^2 L_{\text{fnt}}^2 + n^2\pi^2)$. The total substrate current follows from these carrier profiles: it is the current through the upper depletion region:

$$
\begin{aligned}
J_{\text{subs}}(s) &= eD_p \frac{\partial n_p(s,x)}{\partial x}\Big|_{x=0} \\
&= \sum_{n=1}^{\infty} \frac{2\alpha e\Phi(s)e^{-\alpha(L_x+d)}n^2\pi^2}{L_{\text{fnt}}(\alpha^2 L_{\text{fnt}}^2 + n^2\pi^2)\left(\dfrac{s}{D_n} + \dfrac{1}{L_n^2} + \dfrac{n^2\pi^2}{L_{\text{fnt}}^2}\right)}
\end{aligned}
\qquad (3.15)
$$

The total substrate response is the sum of $n$ Fourier terms with amplitudes that depend on the wavelength. The higher the index $n$, the higher the pole of the corresponding Fourier component, see also figure 3.10. The sum of all components gives the total substrate response with a low roll-off ($\sim$-10 dB/decade).

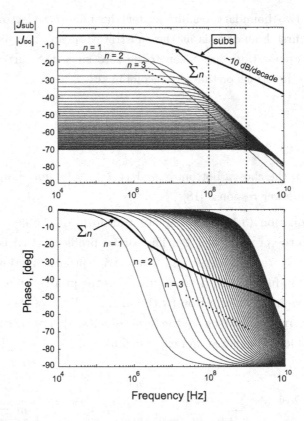

Figure 3.10: Frequency response of the substrate diffusion current: a sum of $n$ one-pole sine-Fourier components.

The calculation of the substrate current response can be simplified by taking an infinite substrate depth (100% of light absorbed), as given in [2]. This solution corresponds to the worst-case solution i.e. with the lowest substrate-current bandwidth. The calculated photocurrent is:

$$J_{\text{subs}}(s) = e\alpha L_n e^{-\alpha L_x} \frac{1}{\sqrt{1 + s\tau_n} + \alpha L_n} \tag{3.16}$$

This equation also shows a *low* current response decay with -10 dB/decade.

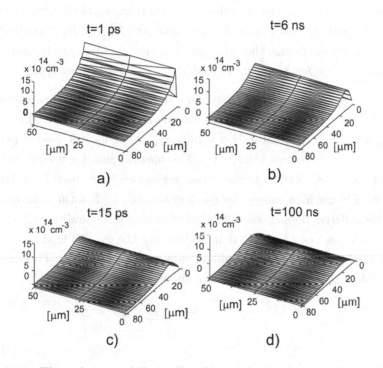

Figure 3.11: The substrate diffusion profile inside p-substrate (depth 80 μm), under incident light pulse (10 ps pulse-width). This profile is shown for 1 ps, 6 ns, 15 ns, and 100 ns.

The inverse Laplace transform of equation (3.15) is used to calculate the substrate current impulse response as a function of time. The result is shown in figure 3.11. The diffusion process is slow and electrons need time (tens of ns) to reach the junctions located close to the photodiode surface. A certain number of carriers will also diffuse deeper into the substrate where they eventually recombines; they do not contribute to the overall photocurrent.

**Depletion region response (drift response)**

The third photocurrent component in equation (3.1), is the drift current inside the depletion regions in the vertical and lateral directions. The drift current is directly related to the depletion volume in which carriers are generated:

$$J_{\text{dcp}} = \Phi e \left( [e^{-\alpha L_x} - e^{-\alpha(L_x+d)}] \frac{A_{\text{total}}}{A_{\text{eff}_{\text{lat}}}} + [1 - e^{-\alpha(L_x)}] \frac{A_{\text{total}}}{A_{\text{eff}_{\text{ver}}}} \right) \qquad (3.17)$$

where $A_{\text{eff}_{\text{lat}}}$ and $A_{\text{eff}_{\text{ver}}}$ are the effective lateral and vertical depletion region areas in comparison with the total photodiode area $A_{\text{total}}$. For twin-well technology with adjoined wells, the side-wall depletion region is much smaller than the bottom one, so for simplicity of calculations it will be neglected.

The velocity of the holes and electrons inside the depletion region depends on the electric field [3]. For very high electric fields ($> 10^7 V$)/m, the speed of both carriers reach their saturation values $v_{n,p_s}$. The electric field depends on the built-in $\phi$ and the bias voltage $V_b$. The bias voltages for nowadays CMOS processes are $\leq 1.8$ V and for the depletion regions width of about $W = 1\mu$m, the electric field is not high enough for carriers to reach their saturation velocities. Due to the different doping concentration of the nwell $N_D$ and substrate regions $N_A$ (the difference can be more than 100 times), the electric field $E(x)$ mainly extends in the region with the lightest doping concentration. The maximum electric field $E_{\text{max}}$ is at the right end of the depletion region [3]:

$$E_{\text{max}} = \frac{eN_A W_{\text{tot}}}{\epsilon_0 \epsilon_r} \qquad (3.18)$$

and

$$E(x) = \xi E_{\text{max}} + \frac{x}{W_{\text{tot}}}(1 - \xi)E_{\text{max}} \qquad (3.19)$$

where $N_A$ is the lighter doping concentration of the substrate, the $\xi$ is the ratio between the minimum electric field (at the beginning of the depletion region) and the maximum electric field, $W_{\text{tot}}$ is the depletion region width, and $x$ is the distance in the depletion region $x \in [0, W_{\text{tot}}]$. The electric field $E(x)$ is shown in figure 3.12.

The transit time of the holes and the electrons inside depletion region is:

$$T_{n,p}(x) = \int_0^{W_{\text{tot}}} \frac{1}{v_{n,p}(x)} dx \qquad (3.20)$$

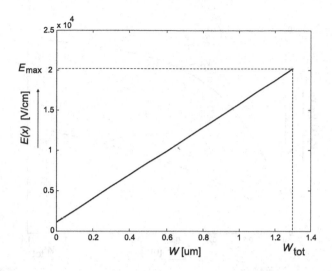

Figure 3.12: A linear approximation of the electric field $E(x)$ inside the depletion region of CMOS photodiode. The minimum electric field is at the beginning of the depletion region, i.e. on the the side with the lowest doping concentration. This electric field is typically insufficiently high for the excess carriers to reach their saturation velocities.

where $v(x)$ is the distance-dependent velocity of the holes and electrons. This velocity can be calculated using the following equations [3]:

$$v_n(E, x) = \frac{v_{sn}}{\left[1 + \left(\dfrac{E_{n0}}{E(x)}\right)^{\gamma_n}\right]^{\frac{1}{\gamma_n}}} \qquad v_p(E, x) = \frac{v_{sp}}{\left[1 + \left(\dfrac{E_{p0}}{E(x)}\right)^{\gamma_p}\right]^{\frac{1}{\gamma_p}}} \qquad (3.21)$$

where $\gamma_n = 2$ and $\gamma_p = 1$, and $v_{sn}$ and $v_{sp}$ are the saturation velocities of the electrons and holes, respectively. These velocities are shown in figure 3.13. Substituting the electric field $E(x)$ with (3.19), and taking the inverse of the velocities results in:

$$\frac{1}{v_p(x)} = \frac{1}{v_{sp}} + \frac{1}{1 + \dfrac{x}{W_{tot}} \dfrac{1-\xi}{\xi}} \frac{E_{p0}}{v_{sp} E_{max}\xi} \qquad (3.22)$$

$$\frac{1}{v_n(x)} = \frac{1}{v_{sn}} \sqrt{1 + \left[\frac{E_{n0}}{\xi E_{max} + \dfrac{x}{W_{tot}}(1-\xi)E_{max}}\right]^2} \qquad (3.23)$$

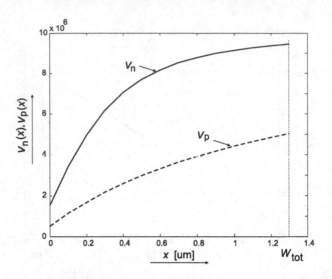

Figure 3.13: The velocity of holes $v_p(x)$ and electrons $v_n(x)$ inside the depletion region under the distance-dependent electric field ($E(x)$ from figure 3.12).

The average transit time of holes and electrons $T_{n,p_{aver}}$ using equations (3.20-3.23) are:

$$T_{n,p_{aver}} = \frac{\int_0^{W_{tot}} T_{n,p}(x) - T_{n,p}(0)dx}{W_{tot}} \tag{3.24}$$

After calculating the average transient time using (3.24), the -3 dB frequency of the holes and electrons are [8]:

$$f_{p,n} = \frac{2.4}{2\pi T_{n,p_{aver}}} \tag{3.25}$$

To calculate the average transit times of the holes and electrons with position dependent electric fields[4], we use the value for the 0.18 $\mu$m CMOS process: $\epsilon_0 = 8.8541 0^{-14}$ F/cm, $\epsilon_r = 11.7$ F/cm, $e = 1.6\cdot10^{-19}$ C, $N_A = 10^{15}$, $N_D = 10^{17}$, $\phi = 0.7$ V, $V_b = -0.8$ V, $\xi = 0.05$. Using equations (3.18-3.25), $T_{p_{aver}} = 3.63\cdot10^{-11}$ s, $f_p = 1.052\cdot10^{10}$ Hz, $T_{n_{aver}} = 8.24\cdot10^{-12}$ s, $f_n = 4.63\cdot10^{10}$ Hz. The frequency response of the depletion region current decays with -10 dB/decade.

---

[4]For most optical receivers, the reverse voltage across the photodiode is large, yielding both a large depletion region width and high (saturated) carrier velocities [8]. In submicron CMOS processes these two are not reached which results in lower performance.

For 0.18 μm CMOS technology, this bandwidth is about $f_{3dBdrift}$=8−10 GHz. These figures are much larger than the diffusion current bandwidth; therefore, for simpler calculations the drift current is taken to be independent on frequency.

### Total photodiode intrinsic characteristics

The sum of all diffusion and drift current components in the previous sections forms the total intrinsic response of the nwell/p-substrate diode. Figure 3.14 shows the calculated responses of the two finger nwell/p-substrate diodes. The first response is for minimal nwell width, which is typically 2 μm for 0.18 μm CMOS. The second response is for an nwell width much larger than its depth, here for 10 μm nwell width. Both responses are calculated for $\lambda = 850$ nm. The values for the parameters in the analytical expressions were directly obtained from the process technology parameters for a fully standard 0.18 μm CMOS process.

For $\lambda = 850$ nm, the substrate current typically dominates the overall photocurrent response up to a few hundreds of MHz. The nwell diffusion current has a larger bandwidth mainly determined by the length of the shortest side of the nwell. For narrow nwells with $L_y = 2$ μm, the shortest sides are both lateral and vertical dimensions. The bandwidth is $f_{3dBnwell} = 930$ MHz. For wide nwell with $L_y = 10$ μm, the shortest side is the nwell depth only, and the charge gradient is lower than in the previous case. The bandwidth in this case is $f_{3dBnwell} = 450$ MHz. Thus, the larger the nwell width $L_y$, in comparison with its depth $L_x$ ($L_y > 2L_x$), the lower the influence of the nwell-width on its bandwidth. The overall maximal intrinsic bandwidth is 5 MHz. This bandwidth is almost independent of the nwell geometry due to the dominant and size-independent substrate current contribution: the fast diffusion response in the nwells and the fast drift response are overshadowed by the large substrate current.

Figure 3.15 shows the physical effects that take place inside a nwell/p-substrate photodiode, after illumination using a Dirac-pulse at $t = 0$ with $\lambda = 850$ nm. The charge generated at $t = 0$ as a function of the depth into the silicon is represented by the upper (continuous) curve. Both the light intensity and the generated charge density decrease exponentially with the depth in the silicon. At 850 nm incident light, the intensity decreases by 50% every 9 $\mu m$, which is much larger than the depth of any junction in standard CMOS technology. For comparison reasons the photodiode structure is sketched on scale below

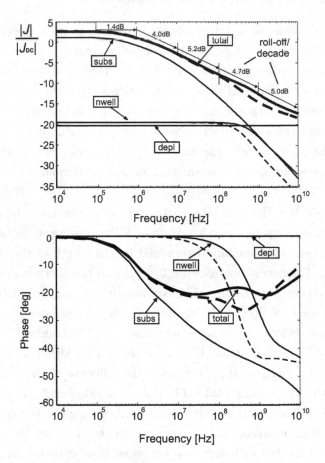

Figure 3.14: The calculated total photocurrent response of nwell/p-substrate photodiode with *high-resistance* substrate in a *twin-well* technology: 2 µm (solid lines) and 10 µm nwell size (dashed lines) for λ = 850 nm.

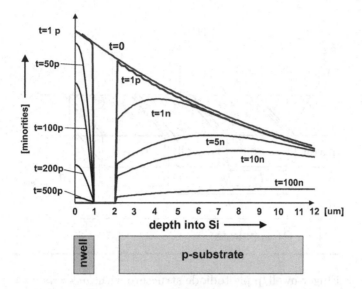

Figure 3.15: Simulated charge distribution in a nwell/p-substrate photodiode after illumination using a Dirac light pulse at t=0, for $\lambda = 850$ nm. The charge profiles at a number of time instances illustrate the speed of response in the time domain in different parts of the photodiode; photodiode dimensions are shown below the graph. Time instances are different for nwell and the p-substrate.

the graph. In figure 3.15, the simulated charge distributions at different time instances illustrate (in the time domain) the fast response of the nwell junction, and much slower response of the charge generated in the p-substrate.

**Roll-off in the frequency characteristics**

The overall intrinsic photodiode response shows a slow decay due to the combination of the three current components. All individual components show a roll-off of -10dB/decade; when summed the roll-off ranges from -4 dB/decade to -10 dB/decade.

The roll-off of the total photocurrent response in the beginning (around the -3 dB point) follows the one of the substrate current response. For higher decades, the total roll-of is smaller in comparison with the substrate roll-of, due to the larger influence of the fast nwell and the depletion region currents. The maximal roll-off value for the frequencies between the -3dB frequency and the lower GHz range is about 5.7 dB/decade for $L_y = 10$ µm and 5.2 dB/decade for $L_y = 2$ µm, as illustrated in figure 3.14. In the low-GHz range, the roll-off

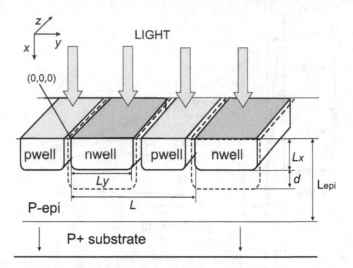

Figure 3.16: Finger nwell/p photodiode structure with *low-resistance* substrate and twin-wells in standard CMOS technology.

is lower (about 4.7 dB/decade) and it decreases with the frequency since the "flat" depletion region response dominates the overall photocurrent.

## Nwell/p-substrate photodiode with low-resistance substrate

In standard CMOS processes, circuit designers can typically choose between "high" or "low"-resistance substrate. This section analyses the photodiode frequency behavior of the nwell/p-substrate with a low-resistance substrate illustrated in figure 3.16.

The only difference between this diode and the previously analyzed photodiode is in the substrate current response. This response is solved again using one-dimensional (vertical) diffusion equation. The two "p" layers are placed at the top of each other (see figure 3.16) and the movement of the minority electrons in both layers is described with two diffusion equations:

$$\begin{aligned}
\frac{\partial n_{\text{p1}}}{\partial t} &= D_{\text{n}} \frac{\partial^2 n_{\text{p1}}}{\partial x^2} - \frac{n_{\text{p1}}}{\tau_{\text{p1}}} + G_1(t,x) \\
\frac{\partial n_{\text{p2}}}{\partial t} &= D_{\text{n}} \frac{\partial^2 n_{\text{p2}}}{\partial x^2} - \frac{n_{\text{p2}}}{\tau_{\text{p2}}} + G_2(t,x)
\end{aligned} \tag{3.26}$$

where the electron generation rate at $t = 0$, in the top substrate layer $G_1(t,x)$ and in the bottom substrate layer $G_2(t,x)$ can be expressed as:

$$
\begin{aligned}
G_1(t,x) &= \alpha\Phi_0(t)e^{-\alpha(L_x+d)}e^{-\alpha x)}\big|_{x\in[0,L_{\text{epi}}]} \\
G_2(t,x) &= \alpha\Phi_0(t)e^{-\alpha L_{\text{epi}}}e^{-\alpha x)}\big|_{x\in[0,\infty]}
\end{aligned}
\tag{3.27}
$$

where $d$ is the depth of the depletion region, and $L_{\text{epi}}$ the depth of the p-epi layer.

In order to calculate the substrate current response in the frequency domain $(s)$, once again the Laplace transform of the diffusion equation (3.26) is taken. Between the two substrate layers there is a boundary condition related to both the current density and the minority carrier concentration [9]. Due to the continuity of currents, the current densities are equal between the two layers:

$$
-qD_{\text{p1}}\frac{\partial n_{\text{p1}}(s,x)}{\partial x}\big|_{x=L_{\text{epi}}} = -qD_{\text{p2}}\frac{\partial n_{\text{p2}}(s,x)}{\partial x}\big|_{x=L_{\text{epi}}}
\tag{3.28}
$$

The second boundary condition is related to the continuity of the concentration of the minority carriers:

$$
n_{\text{p1}}(s,L_{\text{epi}}) = n_{\text{p2}}(s,L_{\text{epi}})
\tag{3.29}
$$

The other two boundary conditions for both electron densities at the bottom of the depletion region, $x = 0$, and at infinite substrate depth, $x = \infty$, are taken to be zero. The infinitely large substrate is taken in order to avoid long and complex calculations [2].

The electrons generated deep in the low-resistance substrate have a higher probability of recombination than in the high-resistance substrate due to the higher doping concentration. The recombined carriers do not contribute to the overall photocurrent. The overall effect of this is that the photo responsivity somewhat decreases, but that at the same time the speed of response increases.

Following the procedure described previously in this section, the total current response is calculated; the result is shown in figure 3.17. In comparison with high-resistance substrate photodiodes, more carriers diffuse towards the substrate bottom resulting in a lower diode DC responsivity. Therefore, the DC current is lower, but the overall bandwidth is higher. The calculated normalized amplitude of the overall photocurrent is 3.5 dB lower but with 2.3 times higher -3 dB frequency: 8 MHz. The photodiode geometry has again almost no

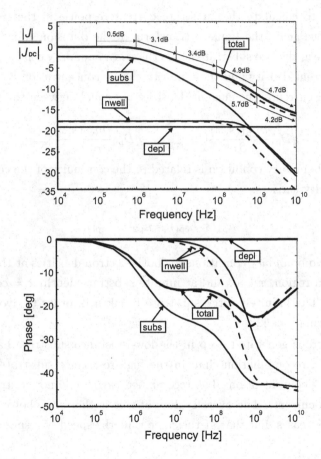

Figure 3.17: The calculated amplitude response of nwell/p-substrate photodiode with *low-resistance substrate* in a *twin-well* technology: 2 µm (solid lines) and 10 µm nwell size (dashed lines) for $\lambda = 850$ nm.

influence on the bandwidth due to the substrate current domination.

The roll-off in the total frequency response for all decades after the -3 dB frequency is **1-2** dB larger (see figure 3.17) in comparison to low-resistance substrate diodes, see figure 3.17.

### 3.2.2 Comparison between simulations and measurements

A finger nwell/p-substrate photodiode with 2 µm nwell width was fabricated in a standard 0.18 µm twin-well CMOS technology. The chip-micrograph is shown in figure 3.18[5]. The overall photodiode area is $50 \times 50$ µm$^2$. The nwells are connected with metal-2 and pwells with metal-1. The total metal area is about 13% of the total photodiode area meaning that the input light signal is decreased for 13%.

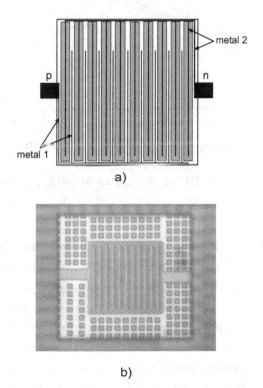

Figure 3.18: a) Layout of nwell/p-substrate photodiode with 2 µm nwell-width in standard CMOS technology b) chip-micrograph.

The responsivity of the photodiode shown in figure 3.18 is measured first

---

[5]The technology has five metal layers and one polyscilicon layer available.

with the DC optical signal. The photoresponsivity and the frequency response are measured using on-chip measurements[6]. The diode is connected to a 1 V DC-supply using a bias-tee. A semiconductor parameter analyzer (SPA) HP4146B is used as the supply. The voltage and current compliances were set in order to avoid incorrect biasing. Using the SPA it was possible to monitor the diode characteristics and to check the correct contacting between the probes and bondpads.

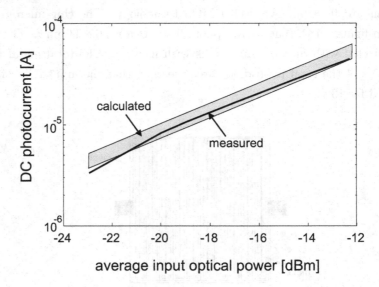

Figure 3.19: The measured DC photocurrent of nwell/p-substrate photodiode for $\lambda = 850$ nm.

The calculated photoresponsivity of the diode, including the metal-coverage and without significant light reflection was 0.56 A/W. The exact amount of re-flection depends on the thickness of the dielectric stack of layers between the sili-con and the air. Best case, these layers are transparent for 850nm light and elim-inate reflections completely by forming an antireflection coating $d = \lambda/4\, n_{\text{top}}$ [3], where $n_{\text{top}}$ is the refractive index of the top transparent layer. Worst case the reflection is not removed and effective amount of optical power incident to a photodiode is $P_{\text{eff}} = 2/3 \cdot P_{\text{in}}$. Due to this uncertainty, the photocurrent values for a different wavelengths can vary by one-third, as shown in figure 3.19. The expected responsivity values range from 0.4 A/W to 0.56 A/W.

An 850-nm VCSEL is used as a light source for measuring the responsivity.

---

[6]The fabricated nwell/p-substrate photodiode was not packaged.

The light was coupled from the laser to the photodiode using a multimode fiber with 50 µm core-diameter; the fiber length is 1 m. The optical power is chosen in the range from 10µW (-20 dBm)[7], up to 120 $\mu$W (-9.2 dBm). The DC optical power at the fiber's output is measured using HP 8153A Lightwave Multimeter. With a responsivity of 0.4 A/W, and 13% of the diode area coveredby metal, the calculated photocurrent is in the range from $I_{photo} = 3.5$µA to 42µA while at 0.56 A/W, the photocurrent is $I_{photo} = 4.9$ µA-59 µA. The measured photocurrent is shown in figure 3.19. This measured current complies with the maximal light reflection case.

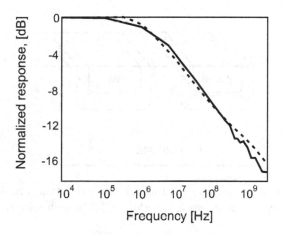

Figure 3.20: The measured (line) and calculated (dashed) responses of a finger nwell/p-substrate photodiode with 2 µm nwell-size, $\lambda = 850$ nm.

The measured frequency response of the photodiode and the calculated response are shown in figure 3.20. For these measurements, the RF-cable response and the laser response are calibrated out for frequencies up to 2 GHz. The measurements are carried out using an E4404E spectrum analyzer. Clearly the measurements comply well to the calculated results.

**Separate wells and high-resistance substrate**

A separate well-CMOS technology combined with a high-resistance substrate is also frequently used [2]; this photodiode structure is shown in figure 3.21. The lateral depletion region between nwells is significantly increased. As a result, the amplitude of the drift current response in the depletion regions is higher

---

[7]-17 dBm sensitivity is specified as the Gigabit Ethernet standard.

in comparison to the adjoined-well diode. The overall -3dB bandwidth remains however about 5 MHz because of the dominant substrate current contribution. On the other side, the depletion region response is dominant in higher decades and the roll-off in the total intrinsic diode characteristic is 1-2 dB lower in comparison with one in the twin well technology. The total intrinsic response of this photodiode in shown in figure 3.22.

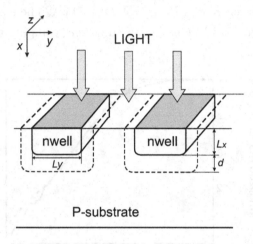

Figure 3.21: A finger nwell/p-substrate photodiode structure with *high resistance* substrate and *separate-wells* standard CMOS technology.

The maximal roll-off value in the photocurrent response is about 5.4 dB/decade for $L_y$=10 μm and 5.0 dB/decade for $L_y$=2 μm, as shown in figure 3.22.

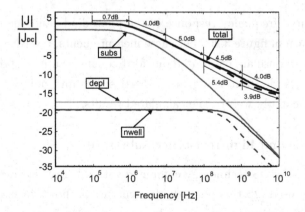

Figure 3.22: The response of nwell/p-substrate photodiode with *high-resistance substrate* in a *separate-wells* technology: 2 μm (solid lines) and 10 μm nwell size (dashed lines) for λ = 850 nm.

### 3.2.3 N+/p-substrate diode

The n+/p-substrate photodiode structure resembles a scaled-down of the nwell/p-substrate junction and it is shown in figure 3.23 . Because these similarities in its construction, the photocurrent response of this diode is similar with the one calculated for the nwell/p-substrate photodiode in the previous sections.

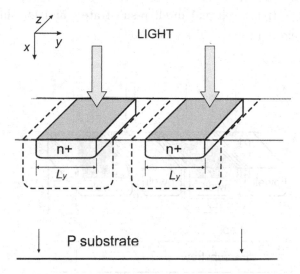

Figure 3.23: Finger n+/p-substrate photodiode with high-resistance substrate.

The response of the $n^+$ region is obtained by replacing both the diffusion length $L_p$ with the $L_{p1}$ and replacing the diffusion coefficient $D_n$ with $D_{n1}$ (corresponding to the doping of the n+ region). Since the doping concentration of the shallow $n^+$ is much higher than that in the n-well, the diffusion length $L_{p1}$ is much smaller [3]. The size of the $n^+$ diffusion layer towards the substrate $L_{x1}$ is also lower. The maximum frequency response is determined mainly by the depth of the n+ region. As a result, the holes' diffusion bandwidth is higher than the one in the nwell region, while the contribution to the overall current response is decreased[8] (see figure 3.23). The depletion region is located closer to the diode surface, which results in the larger drift current (see equations (2.11, 2.12)) , but its influence in the total current response is not changed significantly, because of the still dominant substrate current.

---

[8]The contribution of the slow substrate current is here larger.

### 3.2.4   P+/nwell/p-substrate photodiode with low -resistance substrate in adjoined-well technology

The p+/nwell/p-substrate photodiode consists of two diodes: p+/nwell and nwell/p-substrate, see figure 3.24. The former photodiode can be seen as the complement of the previously discussed n+/p-substrate diode. There are two vertical junctions (p+/nwell and nwell/p-substrate), and for this reason the diode is often referred to as *double photodiode*.

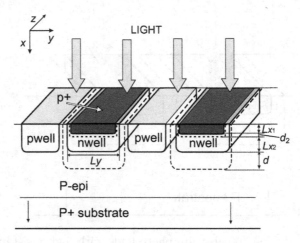

Figure 3.24: Finger p+/nwell/p-substrate photodiode structure in standard CMOS technology with low-resistance substrate and adjoined-wells.

The diffusion current responses are derived using a two-dimensional diffusion equation similar to those in (3.8). The main difference in comparison with the nwell/psubstrate and the n+/p-substrate photodiode analyzed in the previous section is the diffusion response inside the nwell region. The boundary condi-

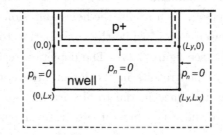

Figure 3.25: The boundary conditions for the hole density inside the nwell for the double-photodiode.

tions for the hole density on every nwell side are shown in figure 3.25; they are
zero since the nwell is enclosed by junctions:

$$p_n|_{x,y@ \text{ boundary}} = 0 \qquad (3.30)$$

The response of the nwell is given in equation (3.31):

$$J_{\text{nwell}_1}(s) = \sum_{n=1}^{\infty} \sum_{m=1}^{\infty} \frac{64e\Phi_0(s)L_p^2[e^{-\alpha(L_{x1}+d_2)} - e^{-\alpha L_{x2}}]}{l\pi^2 L_e}$$

$$\times \frac{\dfrac{L_y}{L_e(2n-1)^2} + \dfrac{L_e}{L_y(2m-1)^2}}{\dfrac{(2n-1)^2\pi^2 L_p^2}{L_y^2} + \dfrac{(2m-1)^2\pi^2 L_p^2}{L_e^2} + s\tau_p + 1} \qquad (3.31)$$

where $L_e = L_{x2} - L_{x1} - d_2$.

For the p+ region , the electron current response is calculated using the nwell
response in the nwell/p-substrate diode using different diffusion coefficients and
diffusion lengths as well as junction depth: ($D_{p1} \rightarrow D_{n1}$, $L_{p1} \rightarrow L_{n1}$, and
$L_x \rightarrow L_{x1}$). The substrate current response for both diodes is the same due to
the same nwell depths and the same doping concentrations.

The total frequency response of the double-photodiode is calculated for two
nwell/p+ sizes: firstly for a narrow nwell, $L_y = 2$ μm, and secondly for a
relatively wide nwell, $L_y = 10$ μm. The wavelength is again $\lambda = 850$ nm.
The results are presented in figure 3.26 showing that the bandwidths of p+
and nwell currents are mainly determined by the low physical depth of the
junctions ($2L_{x1}, 2L_x < L_y$): changing nwell and p+ widths has almost no effect
on the cut-off frequency. The bandwidth of the junction-framed nwell current
is $f_{\text{3dBnwell}} = 5$ GHz for $L_y = 2$ μm, and $f_{\text{3dBnwell}} = 4.2$ GHz for $L_y = 10$
μm. These bandwidth figures are more than twice the nwell bandwidth of the
nwell/p-substrate diode. The distances towards the junctions are lower yielding
a higher charge gradient and higher net transport in the diffusion process. The
current bandwidth of the p+ region is lower than the bandwidth of the nwell
current; the calculated value is about 3 GHz for all nwell/p+ widths, (the depth
of the p+ is smaller than its width, and it mainly determines the bandwidth).
The p+ surface is reflective for the carriers[9]. They are repelled back to the other
three p+ sides with the junctions: they need extra time to start contributing

---

The surface recombination process is slow on the time scales used for signal frequencies
[9]higher than a few Mhz.

to the overall photocurrent. The intrinsic photodiode bandwidth is 4.6 MHz, while the nwell/p+ widths have no significant influence on it (see figure 3.26).

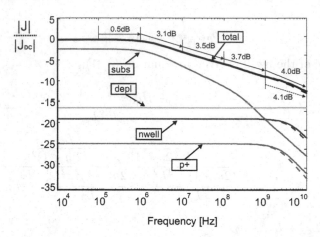

Figure 3.26: The response of *p+/nwell/p-substrate* photodiode with low-resistance substrate in an adjoined-wells technology: 2 µm (solid lines) and 10 µm nwell size (dashed lines) for $\lambda = 850$ nm.

The maximal roll-off value in the photocurrent response is located in the low GHz range and amounts to -4 dB/decade. This holds for both diode geometries, as shown in figure 3.26. In the 1-1000 MHz range, the roll-off per decade is 1-2 dB lower than in nwell/p-substrate diode: the value is about 3.5 dB/decade.

The time impulse response of the hole diffusion profile inside the nwell is again calculated using the Inverse Laplace transform of equation (3.31). This profile is calculated after 1 ps, 6 ps, 15 ps, 100 ps and shown in figure 3.27. The nwell region is completely surrounded by junctions; for this reason the hole-carrier profile diminishes much faster than in the case of the hole profile inside the nwell for the nwell/p-substrate photodiode (see figure 3.8).

Figure 3.28 is a 2D-illustration of the physical effects that take place inside a p+/nwell/p-substrate photodiode, after illumination using a Dirac-pulse at $t = 0$ with $\lambda = 850$ nm. substrate. For illustration purposes the time instances of the charge profiles in p+ and in the nwell are identical; the times in the p-substrate are quite different.

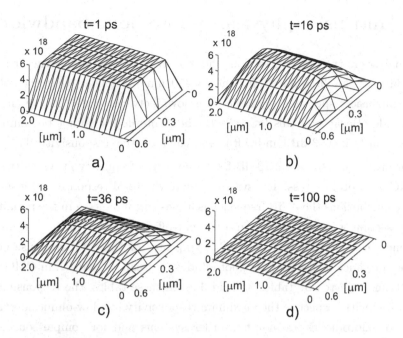

Figure 3.27: The excess carrier concentration in the nwell for the p+/nwell/p-substrate photodiode under incident light pulse (10 ps pulse-width) after 1 ps, 6 ps, 15 ps, 100 ps.

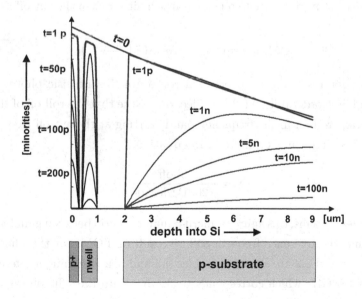

Figure 3.28: Charge distribution in a p+/nwell/p-substrate photodiode after illumination using a Dirac light pulse at t=0, for $\lambda = 850$ nm.

## 3.3   Intrinsic (physical) photodiode bandwidth

The analyses in the previous sections showed various photodiode structures and performances, for 0.18 $\mu$m CMOS at $\lambda$=850 nm. It follows that the roll-off of the intrinsic response of photodiodes is low: between -3 dB/decade and -10 dB/decade. Because of this low roll-off, the photodiodes cannot be compared based on just their -3 dB bandwidths and their relative responsivity [11].

For ordinary systems the -3 dB frequency is the frequency at which the DC and AC asymptotes cross. For systems where the total response is the sum of many contributions the -3 dB frequency is almost meaningless. In the remainder of this section we will therefore use the cut-off frequency, which is again the frequency at which the DC and AC asymptotes of the total response cross. The performance of a number of photodiodes that can be realized in CMOS for $\lambda$=850 nm are listed in Tables 3.1 and 3.2. In these tables, the responsivity is normalized with respect to the maximum responsivity in a low-ohmic substrate. All these photodiodes are non-first-order systems and for comparison of the performance of the intrinsic performance of photodiodes a new figure of merit (FOM$_i$) is introduced. In analogy to the gain-bandwidth product in amplifiers, a good FOM$_i$ is the responsitivy at a certain reference frequency resp($f_{ref}$). Assuming that this reference frequency is much higher than the cut-off frequency $f_{cut_{off}}$:

$$\text{FOM}_i = \text{resp}(f_{ref}) = \text{resp}(0) \left( \frac{f_{cut_{off}}}{f_{ref}} \right)^s \tag{3.32}$$

where the factor $s$ is the ratio between roll-off of the intrinsic photodiode response and first-order roll-off (-20 dB/decade). Note that the roll-off of the diode is the average roll-off in the frequency band starting at the cut-off frequency up to the highest frequency of interest. In equation,

$$s = \frac{\text{rolloff}}{-20\text{dB/decade}} \tag{3.33}$$

For first order systems, this figure-of-merit equals the ratio between gain-bandwidth product and the reference frequency. The resulting FOMs for the photodiodes in Tables 3.1 and 3.2 are shown in Tables 3.3 and 3.4, assuming a reference frequency of 1.5 GHz which corresponds to 3 Gb/s data-rate. It follows that the photodiodes on low-ohmic substrates have the highest performance. Furthermore, narrow finger structures perform a little better than wide finger structures although the impact of layout optimization is not significant at $\lambda$=850 nm. The best structure is clearly the complex p+/nwell/p-substrate photodiode; the second best is the simpler nwell/p-substrate one.

Table 3.1: The cut-off frequency and responsivity of nwell/p-substrate photodiode with various substrates

| | $\lambda = 850\text{nm}$ | |
| --- | --- | --- |
| | $L_y = 2\mu\text{m}$ | $L_y = 10\mu\text{m}$ |
| high-resistance substrate separate-wells | | |
| cut-off freq | 1MHz | 1MHz |
| average roll-of/decade | 4.6 | 4.9 |
| normalized responsivity | 3dB | 3dB |
| low-resistance substrate separate-wells | | |
| cut-off freq | 1.4MHz | 1.4MHz |
| average roll-of/decade | 3.9 | 4.2 |
| normalized responsivity | 0dB | 0dB |
| high-resistance substrate adjoined-wells | | |
| cut-off freq | 0.6MHz | 0.6MHz |
| average roll-of/decade | 4.3 | 4.8 |
| normalized responsivity | 3dB | 3dB |
| low-resistance substrate adjoined-wells | | |
| cut-off freq | 1MHz | 1MHz |
| average roll-of/decade | 3.9 | 4.4 |
| normalized responsivity | 0dB | 0dB |

## 3.4 Extrinsic (electrical) photodiode bandwidth

Apart from the intrinsic bandwidth of the "stand-alone" photodiode, the in-circuit photodiode bandwidth is also determined by the extrinsic (electrical) bandwidth. This bandwidth is determined by the diode and interconnect capacitance in combination with the pre-amplifier's input resistance.

Table 3.2: The cut-off frequency and responsivity of p+/nwell/p-substrate photodiode with various substrates

|  | $\lambda = 850$nm | |
| --- | --- | --- |
|  | $L_y = 2\mu$m | $L_y = 10\mu$m |
| high-resistance substrate separate-well | | |
| cut-off freq | 1.6MHz | 1.6MHz |
| average roll-of/decade | 3.9 | 3.9 |
| normalized responsivity | 3dB | 3dB |
| low-resistance substrate separate-well | | |
| cut-off freq | 2.0MHz | 2.0MHz |
| average roll-of/decade | 3.3 | 3.3 |
| normalized responsivity | 0dB | 0dB |
| high-resistance substrate adjoined-well | | |
| cut-off freq | 1.2MHz | 1.2MHz |
| average roll-of/decade | 3.9 | 4.0 |
| normalized responsivity | 3dB | 3dB |
| low-resistance substrate adjoined-well | | |
| cut-off freq | 1.8MHz | 1.8MHz |
| average roll-of/decade | 3.1 | 3.15 |
| normalized responsivity | 0dB | 0dB |

The capacitance of high-speed photodiodes depends on the diode area and it is typically in the pF range (using a multimode fiber connection the diode area is $50\times50$ µm$^2$). Table 3.5 shows the calculated values of the parasitic capacitances for two nwell widths of nwell/p-substrate and double photodiode in the 0.18 µm CMOS technology: the first nwell width is twice its depth $L_y = 2L_x = 2$ µm, and the second with is much higher than the nwell depth $L_y = 10$ µm.

For the photodiodes in the separate-wells technology, the width of the lateral depletion region is much larger than for the diodes in twin-well technology with

Table 3.3: The FOM of the intrinsic performance of the nwell/p-substrate photodiode for $f_{ref} = 1.5$ GHz.

| | $\lambda = 850$nm | |
|---|---|---|
| | $L_y = 2\mu$m | $L_y = 10\mu$m |
| high-resistance substrate separate-wells | | |
| FOM$_i$ | 0.73 | 0.59 |
| low-resistance substrate separate-wells | | |
| FOM$_i$ | 0.77 | 0.69 |
| high-resistance substrate adjoined-wells | | |
| FOM$_i$ | 0.63 | 0.5 |
| low-resistance substrate adjoined-wells | | |
| FOM$_i$ | 0.68 | 0.57 |

Table 3.4: The FOM of the intrinsic performance of the p+/nwell/p-substrate photodiode for $f_{ref} = 1.5$ GHz.

| | $\lambda = 850$nm | |
|---|---|---|
| | $L_y = 2\mu$m | $L_y = 10\mu$m |
| high-resistance substrate separate-well | | |
| FOM$_i$ | 1.05 | 1.05 |
| low-resistance substrate separate-well | | |
| FOM$_i$ | 0.98 | 0.98 |
| high-resistance substrate adjoined-well | | |
| FOM$_i$ | 0.998 | 0.998 |
| low-resistance substrate adjoined-well | | |
| FOM$_i$ | 1 | 1 |

adjoined wells. The doping concentration of the pwells is about two orders of
magnitude larger than in high-resistance substrate. Hence, the calculated depletion region width towards pwells is 7 times smaller. The total diode capacitance
is 5-7 times larger for a adjoined-well process in comparison with separate-well
processes.

Table 3.5: Parasitic capacitance for different photodiode structures and geometries

| | $L_y = 2$ $\mu$m | $L_y = 10$ $\mu$m | FOM$_{ex_2}$ | FOM$_{ex_{10}}$ |
|---|---|---|---|---|
| nwell/p-substrate | | | | |
| separate-wells | 0.28 pF | 0.27 pF | 189$\Omega$ | 196$\Omega$ |
| adjoined-wells | 1.6 pF | 0.62 pF | 33$\Omega$ | 85$\Omega$ |
| p+/nwell/p-substrate | | | | |
| separate-wells | 2.0 pF | 1.8 pF | 26$\Omega$ | 29$\Omega$ |
| adjoined-wells | 3.60 pF | 2.20 pF | 15$\Omega$ | 24$\Omega$ |

For all photodiodes discussed in this chapter, there are two general observations for their electrical bandwidth. First, by decreasing the diode nwell width
(for a constant diode area), the total junction area of the photodiode increases.
As a result, the diode capacitance increases. Second, the implementation of
twin-well technology increases diode capacitance too. This implementation par-
ticularly changes the nwell/p-substrate diode capacitance.

From a circuit point of view, a reasonable FOM for the extrinsic behavior of
photodiodes, FOM$_{ex}$, is the required input resistance to reach a certain bandwidth, assuming an ideal preamplifier with a purely capacitive input impedance.
This FOM is proportional to the ease of implementing a suitable pre-amplifier
input stage for the photodiode. The FOM$_{ex}$ calculated for an electrical bandwidth of 3 GHz are also shown in Table 3.5.

Combining the FOMs for the intrinsic and extrinsic performance of CMOS
photodiodes, shown in Tables 3.3-3.4 and Table 3.5, it follows that there is
no such thing as best photodiode based on only photodiode properties. The
selection of the best photodiode for our application hence includes system and
circuit aspects; the selection is done in chapter 4.

## 3.5 Noise in photodiodes

The noise generated by a photodiode operating under reverse bias, is a combination of shot noise and Johnson noise. Shot noise is generated by random fluctuations of current flowing through the device. This noise is discovered in tubes in 1918 by Walter Schottky who associated this noise with direct current flow. The dc current is a combination of dark current ($I_r$) and quantum noise ($I_{qn}$). Quantum noise results from generation of electrons by the incident optical radiation. The shot noise is given as [3]:

$$\overline{i_s^2} = 2q(I_r + I_{qn}) \cdot BW \tag{3.34}$$

where $\overline{i_s^2}$ is the shot-noise current, and $BW$ is bandwidth of interest.

## 3.6 Summary and conclusions

This chapter analyzed the frequency and the time responses of different photodiode structures in a standard 0.18 $\mu$m CMOS technology, for $\lambda = 850$ nm. The photodiodes are first analyzed as stand-alone detectors i.e. without subsequent electronic circuitry. This allows an analysis of the *intrinsic* photodiode behavior. The intrinsic behavior is related to the movement (drift and diffusion) of the generated carriers inside the diode. Based on the frequency analysis, an intrinsic (physical) diode bandwidth is determined and an intrinsic FOM$_i$ is introduced. In the second part of this chapter, the diode is investigated as an "in-circuit" element, integrated together with the subsequent electronics. The *electrical* bandwidth of the diode is determined by the diode capacitance and the input impedance of the subsequent amplifier. The ease of implementation of a TIA can be estimated using a novel FOM$_{ex}$ indication.

# Bibliography

[1] IEEE 10 Gigabit Ethernet Standard 802.3ae.

[2] D. Coppée, H. J. Stiens, R. A. Vounckx, M. Kuijk: "Calculation of the current response of the spatially modulated light CMOS detectors", *IEEE Transactions Electron Devices*, vol. 48, No. 9, 2001, pp. 1892-1902.

[3] S. M. Sze: *"Physics of semiconductor devices"*, New-York: Wiley-Interscience, 2-nd edition, 1981.

[4] G.W. de Jong et al.: "A DC-to-250MHz Current Pre-Amplifier with Integrated Photo-Diodes in Standard CBiMOS, for Optical-Storage Systems", *ISSCC 2002*, s21.8

[5] S. Radovanović, A. J. Annema and B. Nauta: "Physical and electrical bandwidths of integrated photodiodes in standard CMOS technology", *Electron Device Solid State Circuit 2003*, Hong Kong, pp. 95-98.

[6] S. Radovanović, A. J. Annema and B. Nauta: "On optimal structure and geometry of high-speed integrated photodiodes in a standard CMOS technology", *CLEO 2003*, Taiwan, pp.87, TU4H-9-1.

[7] C. L. Schow, J. D. Schaub, R. Li, and J. C. Campbell, "A 1 Gbit/s monolithically integrated silicon nmos optical receiver", *IEEE J. Select. Topics Quantum Electron.*, vol. 4, pp. 1035-1039, Nov.Dec. 1999.

[8] S. Alexander: *"Optical communication receiver design"*, SPIE Optical engineering press, 1997.

[9] W. J. Liu, O. T.-C. Chen, L.K. Dai and F. W. J. C. Cheng: "A CMOS Photodiode Model", *2001 IEEE International Workshop on Behavioral Modeling and Simulation*, Santa Rosa, California, October 10-12, 2001.

[10] Y. R. Nosov: *"Switching in semiconductor diodes"*, New York: Plenum, 1969, p.14.

[11] S. Radovanovic, A.J. Annema and B. Nauta, "A 3-Gb/s Optical Detector in Standard CMOS for 850-nm Optical Communications", *IEEE Journal of Solid-State Circuits*, vol. 40, August 2005

# CHAPTER 4

# High data-rates with CMOS photodiodes

*The speed of photodiodes in standard CMOS is low. This chapter presents a circuit approach that enables high data-rates, even using the slow CMOS photodiodes. The solution presented is an inherently robust analog equalizer that exploits the properties of CMOS photodiodes to the maximum.*

## 4.1 Introduction

The intrinsic, physical, bandwidth of photodiodes in standard deep-submicron CMOS technologies is around 1MHz for $\lambda = 850$ nm. Assuming an ideal transimpedance amplifier this intrinsic frequency response of the photodiode is its overall response.

It is well known that high inter symbol interference (ISI) levels occur if the bit rate is much higher than the bandwidth of the used channel [1]. High levels of ISI, in turn, result in high bit error rates (BER), see e.g. section 4.2.2. It can be concluded that standard CMOS photodiode at $\lambda = 850$ nm cannot be used straight-forwardly to get bit rates much higher than a few Mb/s.

To be able to operate CMOS photodiodes on high data-rates (at least hundreds of Mb/s) at $\lambda = 850$ nm, the ISI at high frequencies must be reduced significantly. In literature, the most common solution for reducing ISI is the application of an adaptive equalization, either in the analog or in the digital domain. Equalization has been widely used in communications applications such as voice-band modems, wireless [2], digital subscriber lines, and ISDN [3], and even at rates close to 500 Mb/s in disk drives [4, 5]. In all of these applications the equalizer corrects for the channel. Also for long-haul fiber optics communication, adaptive equalization is typically used for fiber dispersion compensation [6]. In this chapter equalization is used to compensate for the intrinsic photodiode response: for imperfections in the receiver itself.

As design vehicle, an optical detector system targeting at 3 Gb/s is used in this chapter [7, 8]. With the assumption that the electrical time constant is sufficiently small, the equalization of the intrinsic photodiode bandwidth must be at least up to 1 GHz to get a sufficiently low ISI[1]. The input impedance of the pre-amplifier is designed to give an electrical bandwidth significantly higher than this equalization range. This electrical bandwidth is maximized by minimizing both the input resistance of the subsequent pre-amplifier and the photodiode capacitance. Note that minimizing input resistance if a pre-amplifier typically increases the power consumption of the pre-amplifier.

The resulting system setup of the optical detector is is shown in figure 4.1. The difference in comparison with the straightforward pre-amplifier configuration in the optical receiver, e.g. in [9], is that an equalizer is placed after the transimpedance amplifier (TIA). In this manner the signal-to-noise ratio (SNR) is maximized [6]. In the following sections, the various parts of the system are worked out in detail. Section 4.2 briefly reviews some design aspects that are relevant for the trans impedance amplfier; these aspects include noise and bandwidth limitations. In section 4.3 the "best" photodiode is selected. This selection procedure includes photodiode properties, circuit properties and performance targets. The actual design of the equalizer is presented in secion 4.4, while robustness aspects are analyzed in 4.5. Measurements on the total designed system are presented in section 4.6.

---

[1]A higher bandwidth results in lower ISI and hence better sensitivity or lower BER, but comes at the cost of a higher power consumption for the pre-amplifier. The 1 GHz bandwidth is sufficient to reach sufficiently low BER figures, at near minimum power consumption at 3 Gb/s.

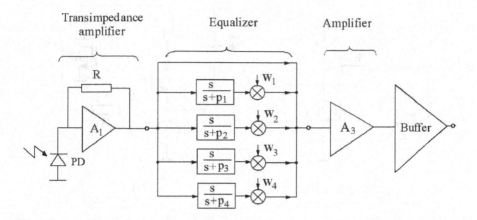

Figure 4.1: Block-diagram of integrated photodiode and preamplifier system using an equalizer to compensate for the photodiode's response.

## 4.2 Transimpedance amplifier design

Transimpedance amplifiers(TIA) are typically used as current-to-voltage convertor for optical receivers. Their use is to increase the bandwidth by providing a low impedance input,and converting the input signal (current) into a voltage. Typically, during the TIA design, the main tradeoffs are in sensitivity (due to noise), speed (bandwidth) and transimpedance gain. The transimpedance gain is typically equal to the feedback resistor for large open-circuit amplifications [10]. If the output voltage signal is small, further amplification is done by a post-amplifier. A large feedback resistance increases the gain, but at the same time may reduce the amplifiers' bandwidth [10].

In general, there are two basic transistor configurations for a TIA design: common source (CS) and common gate (CG) [10, 11]. These two are shown in figure 4.2. The implementation of one of the two configurations depends on the TIA performances: required transimpedance, bandwidth, noise, power consumption.

In this work, the main issue in the pre-amplifier design was to demonstrate the effects of the equalization to robustly compensate for the photodiode response. Because of this, a relatively simple single-stage common source TIA configuration was chosen. To get sufficient overall transimpedance, a number of gain stages are added.

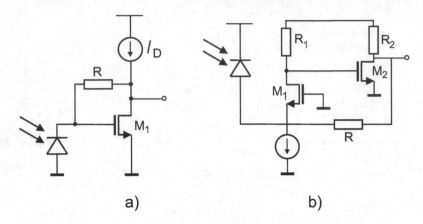

a)                              b)

Figure 4.2: Low input impedance transimpedance amplifiers a) common source, b) common gate amplifier.

## 4.2.1   Transimpedance amplifiers and extrinsic bandwidth

The electrical bandwidth of a photodiode-TIA system is usually determined by the pole at the input of the TIA, formed by the total input capacitance seen at the input node and by the input resistance [10]. Typically the demand on input resistance translates directly in a lower bound on the bias current in the input stage. For the input stage shown in figure 4.2, the input resistance equals:

$$r_{\text{in}} = \frac{R_0 + R}{1 + g_m R_0} \approx \frac{1}{g_m} \qquad (4.1)$$

where   $g_m$ is the transistor's transconductance

$R_0$ is the resistive load at the output nodes

At constant effective gate-source overdrive voltage $V_{\text{gs}} - V_T$ it can easily be shown that

$$r_{\text{in}} \approx \frac{1}{g_m} \propto \frac{1}{I_D} \propto \frac{1}{W} \qquad (\text{constant} V_{\text{gs}} - V_T) \qquad (4.2)$$

This input resistance can be decreased by simultaneously increasing the transistor width $W$ and its bias current $I_D$. If the diode capacitance is dominant in the total capacitance at the input of the TIA, the capacitance at the input node is $C_{\text{in}} \approx C_{\text{diode}}$. As a result, then the required input resistance is $r_{in} \approx$

$\text{FOM}_{\text{ex}}$. Note that equation (4.2) also shows that the bias current of the input stage $I_D$ is inversely proportional to the extrinsic $\text{FOM}_{\text{ex}}$ shown in Table 3.5. In a realistic case, the *total* capacitance at the input node of the TIA must be taken into account. Noting that the TIA itself adds to this total capacitance, $C_{in} > C_{\text{diode}}$; the required input resistance is then $r_{in} < \text{FOM}_{\text{ex}}$.

## 4.2.2 Impact of noise: BER

For high-speed data-communication, the achievable data-rates are closely linked to proper bit detection. The measure for the data quality is the bit error rate (BER). Today's optical links requires BER$\leq 10^{-11}$ [12].

It was shown in e.g. [1] that ISI and BER are closely related. Considering binary data at the transmitter side and a fixed threshold level at the detector side (typically half the output value), this ISI-BER relation is

$$\text{BER} = \frac{1}{2}\text{erfc}\left(\frac{S}{\sqrt{ISI_v^2 + N^2}}\right) \tag{4.3}$$

$$\begin{aligned}\text{with} \quad &ISI_v \quad \text{the statistical variance of ISI}\\ &S \quad \text{the rms signal value}\\ &N \quad \text{the rms noise value}\end{aligned}$$

Clearly, apart from the ISI component, the BER relation includes a signal (S) and a noise (N) term. The S-term is simply the rms value of the received bit symbol. It was derived in [1] that the rms signal for a bit-period $T_b$ is

$$S = \int_0^\infty J(t)[H(t) - H(t - T_b)]dt \tag{4.4}$$

where $H$ denotes Heaviside function. The time domain current impulse response $J(t)$ can be obtained from the inverse Laplace transform of the frequency response of the photocurrent. Figure 4.3 shows the random bitstream BER of the equalized signal for several signal-to-noise levels. The data-rate is normalized to the electrical bandwidth of the total system. Note that for a normalized data-rate lower than about 1.5 b/Hz the BER is noise-limited.

Figure 4.3 shows that there are many combinations of normalized data-rate and SNR leading to a certain BER value. This makes it possible to trade power efficiency issues for noise against those for speed. For our design vehicle, targeting at 3 Gb/s data rates at BER$=10^{-12}$, we selected an electrical bandwidth of

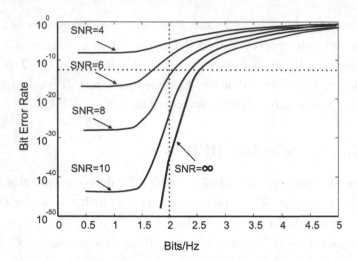

Figure 4.3: BER as a function of the normalized data-rate for several signal-to-noise ratios. The data-rate is normalized to the electrical bandwidth of the system.

1.5 GHz and a signal-to-noise ratio $S/N \approx 8$ which yields a low overall power consumption. This bandwidth is marked by the vertical line in figure 4.3, while the BER value corresponds to the horizontal one. The SNR demand follows from at the crossings of these two lines.

### 4.2.3   Noise of the TIA

It follows from the analyses in chapter 3 that for the CMOS photodiode with the highest responsivity, at $\lambda = 850$ nm and with an active area of about $50\mu m \times 50\mu m$, the rms value of the photocurrent is about 5 $\mu A$. For a $S/N = 8$ then the rms value of the input-referred noise is $i_n = 0.63$ $\mu A$. For the total optical receiver circuit, we used 6 stages, all contributing about equally to the total noise. Taking into account the gain throughout the receiver, this means that the demands on the first stage are the hardest to meet. This first stage also must satisfy input resistance demands to get sufficient bandwidth.

For a common-source TIA configuration, shown in figure 4.4, the output signal and the output noise can easily be derived frequencies lower than the circuit's bandwidth. The various properties of the first stage are:

$$\frac{v_{out,s}}{i_{photo}} = R_o \cdot \frac{1 - R \cdot g_{m1}}{1 + R_o \cdot g_{m1}} \approx -R$$

$$r_{in} = \frac{R + R_o}{1 + R_o \cdot g_{m1}}$$

$$r_{out} = \frac{R_o}{1 + R_o \cdot g_{m1}} \approx \frac{1}{g_{m1}}$$

$$\frac{v_{out,n}^2}{BW} \approx 4kTR + \frac{8\,kT\,(g_{m1} + g_{m2})}{3} \frac{1}{g_{m1}^2} \qquad (4.5)$$

with $R_o$ the combined output resistances of $M_1$ and $M_2$

with $g_{mn}$ the transconductance of $M_n$

To get an overall -3dB bandwidth of about 1.5 GHz for the 6 stages, the bandwidths of all the individual stages are roughly 4.3 GHz. This last figure accounts for a noise-bandwidth of about 5 GHz. Note that with these assumptions both lower noise and lower input impedance can be obtained at the cost of power consumption.

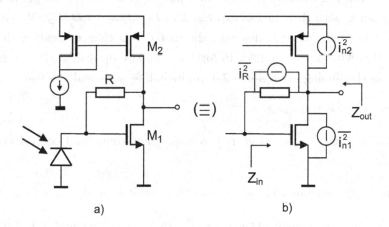

a)                    b)

Figure 4.4: The simplified noise model of the CS TIA.

For our circuit topology and with the previous assumptions, a simple estimation can be made of the dominant effect in the overall power consumption. Three observations can readily be made:

- at low power consumption the input resistance is high yielding a too low bandwidth for photodiodes with a low $FOM_i$.

- at low power consumption the noise level is too low to get an acceptable BER.

- at increasing power consumption levels both the bandwidth increases and the the noise level decreases.

We now distinguish two cases. Firstly for a number of photodiodes that have a high $FOM_i$ the demand on sufficiently low input impedance is more easily met than that for low noise level. For these photodiodes the *noise demands determine the minimum power level*. In the same way, for diodes with a low $FOM_i$ the noise demands are easier to reach than the low input impedance aspect: for these photodiodes *the power consumption is determined by the demands on low input impedance*. This classification is used in the next section to select the most suitable photodiode (using the assumptions made).

## 4.3  Photodiode selection

In chapter 3 a full discussion of the various properties of CMOS photodiodes was presented, with their fitness expressed in $FOM_i$ and $FOM_{ex}$. Whereas in chapter 3 the bare photodiodes were discussed, this chapter deals with their combination with a pre-amplifier to form a complete optical receiver frontend. Extending the finding of section 4.2.3, by including non-idealities like:

- excess noise of transistors,

- the low effective gate-overdrive voltages that come for free in deep sub-micron processes,

- the significant input capacitance of the circuit formed,

a classification can be made of the photodiodes discussed in chapter 3. Table 4.1 first lists the intrinsic FOM$_i$ and the extrinsic FOM$_{ex}$ of all previously discussed CMOS photodiodes. As discussed in section 4.2.3 for some photodiodes the noise-demands determine the power consumption while for others the input resistance demands do. The fourth column in table 4.1 shows which effect is dominant[2] for the power consumption for each photodiode.

For this work, the available 0.18 $\mu$m CMOS technology has a low-resistance substrate and adjoined wells. Because of these (practical) reasons, photodiode

---

[2]With the assumptions in section 4.2.3, and targeting at a 3 Gb/s data rate.

Table 4.1: The FOM and the dominant effect for the input stage's bias current for different photodiode structures and geometries.

| type | | $FOM_i$ | $FOM_{ex}$ | dominant for $I_D$ |
|------|--------------------------------|---------|--------------|--------------------|
| A | p+/nwell/p-subs separate-wells | | | |
| | 2 $\mu$m nwell | 1.05 | 26 $\Omega$ | $r_{in}$ |
| | 10 $\mu$m nwell | 1.05 | 29 $\Omega$ | $r_{in}$ |
| B | p+/nwell/p-subs adjoined-wells | | | |
| | 2 $\mu$m nwell | 1 | 15 $\Omega$ | $r_{in}$ |
| | 10 $\mu$m nwell | 1 | 24 $\Omega$ | $r_{in}$ |
| C | nwell/p-subs separate-wells high-resistance substrate | | | |
| | 2 $\mu$m nwell | 0.73 | 189 $\Omega$ | SNR |
| | 10 $\mu$m nwell | 0.59 | 196 $\Omega$ | SNR |
| D | nwell/p-subs adjoined-wells high-resistance substrate | | | |
| | 2 $\mu$m nwell | 0.63 | 33 $\Omega$ | SNR |
| | 10 $\mu$m nwell | 0.5 | 85 $\Omega$ | SNR |
| E | nwell/p-subs separate-wells low-resistance substrate | | | |
| | 2 $\mu$m nwell | 0.77 | 189 $\Omega$ | SNR |
| | 10 $\mu$m nwell | 0.69 | 196 $\Omega$ | SNR |
| F | nwell/p-subs adjoined-wells low-resistance substrate | | | |
| | 2 $\mu$m nwell | 0.68 | 33 $\Omega$ | SNR |
| | 10 $\mu$m nwell | 0.57 | 85 $\Omega$ | SNR |

types A, C, D and E in table 4.1are discarded in this chapter. Of the remaining two types, B has better intrinsic performance but its power consumption is limited by the requirements on $r_{in}$. Type F has a somewhat worse intrinsic behavior with a lower, noise limited, power consumption. It appears that the narrow-finger nwell/p-substrate photodiode (F) yields the best performance for low-power applications targeting at 3 Gb/s data-rates.

## 4.4    Equalizer design

It was shown in chapter 3 that the total frequency response of the photodiode is the sum of the frequency responses of particular diode regions, resulting in a low roll-off. The equalization characteristics in this book is the complement of the frequency characteristics of the implemented photodiode. One way to mimic a low roll-up characteristics is summation of the outputs of four parallel first-order high-pass filters (HPF); this is illustrated in figure 4.1. The equalizer characteristic is shown in figure 4.5.

The number of high-pass sections is based on the required equalization accuracy: less sections give a too coarse equalization while 4 sections appears to be sufficient. More sections could be used, resulting in a slight increase in performance at the cost of power and area consumption. Many more section are useless due to component spread.

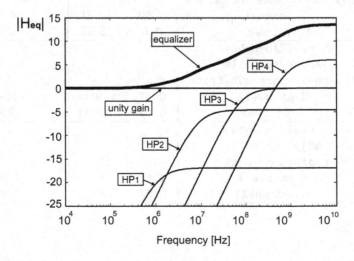

Figure 4.5: The characteristics of the analog equalizer of figure 4.1: the sum of unity gain path and 4 high-pass sections.

One way to realize the analog equalizer is to use a source degeneration (SD) stage with low-pass filter sections in its source [6]; this configuration is shown in figure 4.6. Assuming a high $g_m$ for transistor $M_S$ and with $R_D = R_S$, the transfer function $V_{\text{out}}/V_{\text{in}}$ of the equalizer can be approximated by:

$$\frac{V_{\text{out}}}{V_{\text{in}}} \approx - \left( 1 + \frac{sR_DC_1}{1+sR_1C_1} + \frac{sR_DC_2}{1+sR_2C_2} + \frac{sR_DC_3}{1+sR_3C_3} + \frac{sR_DC_4}{1+sR_4C_4} \right) \quad (4.6)$$

The magnitude of this transfer function is

$$\left| \frac{v_{out}}{v_{in}} \right| = 1 \qquad \text{at low frequencies}$$

$$\left| \frac{v_{out}}{v_{in}} \right| = 1 + \frac{R_D}{R_1} + \frac{R_D}{R_2} + \frac{R_D}{R_3} + \frac{R_D}{R_4} \qquad \text{at high frequencies}$$

with a low roll-up behavior for intermediate frequencies.

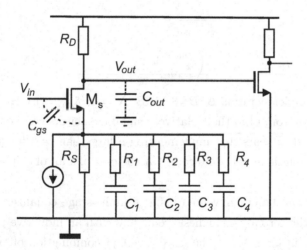

Figure 4.6: The analog equalizer from figure 4.7 including parasitic capacitances.

The output pole of the SD stage itself is determined by the total capacitance at the drain node of $M_S$, $C_{\text{out}}$, and by the resistance at this node, $R_D$: $f_p = 1/2\pi C_{\text{out}} R_D$. For proper operation of the equalizer this pole should be high enough not to interfere with the equalization range: $f_p > f_{3\text{dB}_{\text{elec}}}$.

In our design the proceeding common-source stage is dominant in the $C_{\text{out}}$. In this case lower values for $R_D$ increase the bandwidth of the SD stage. It follows from (4.6) that for proper equalization characteristics, low values of $R_D$

require both low values for $R_{1...4}$ and proportionally higher values for $C_{1...4}$. Clearly for the circuit in figure 4.6 there is a trade-off between bandwidth of the equalizer and area consumption.

The diode frequency characteristics of the narrow-finger nwell/p-substrate photodiode is shown in figure 3.17. This response roughly drops 15 dB in the frequency range from 1 MHz to 1 GHz. A first order estimation for the equalizer is then:

- 4 logarithmically spaced time constants

- about 3.75 dB gain per stage

In equation this yields:

$$R_n = \frac{R_D}{10^{\frac{\Delta dB \cdot n}{20}} - 10^{\frac{\Delta dB \cdot (n-1)}{20}}} \tag{4.7}$$

$$C_n = \frac{1}{2\pi \cdot f_n \cdot R_n}$$

$$f_n = f_{min} \cdot \left(\frac{f_{max}}{f_{min}}\right)^{\frac{n-1}{N}} \qquad n = 1...N$$

while for our implementation $\Delta dB$=3.75dB, $N = 4$, $f_{max}$ =1 GHz and $f_{min}$ =1 MHz. It follows from (4.8) that relatively small capacitors can be used for high values of $R_D$. It appears that implementing a large bandwidth equalizer using the circuit schematic in figure 4.6 comes at the cost of a lot of chip area or power consumption.

This trade-off can be circumvented using a multi-stage equalizer. For this we implemented the zero at the highest frequency with an inductive peaking stage [14], as shown in figure 4.7. The $g_m(C_{gs} + C_1)$ combination of transistor $M_p$ together with resistor $R_1$, behaves as an inductor in a certain frequency range. The impedance at the drain of transistor $M_3$ is:

$$Z_d = \frac{1}{g_{m_{Mp}}} \frac{1 + sR(C_{gs} + C_1)}{1 + s(C_{gs} + C_1)/g_{m_{Mp}}} \tag{4.8}$$

where $R > 1/g_m$. In this manner the circuit with the analog equalizer with one inductive peaking stage and the source degeneration with the first three poles of the equalizer is designed. The overall circuit is shown in figure 4.7. The simulated frequency response at the pre-amplifier's output, including the photodiode, is shown in figure 4.8. The equalizer's frequency response is band-

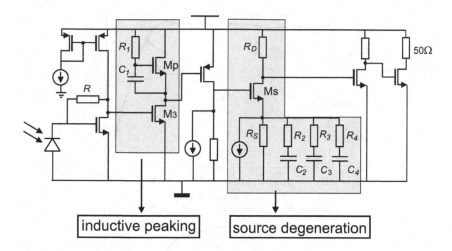

Figure 4.7: The circuit topology of the preamplifier including the analog equalizer.

limited to prevent out of band high-frequency noise from being added to the signal [15].

## 4.5 Robustness on spread and temperature

For any equalizer system, robustness aspects against non-idealities such as spread and temperature fluctuations is of major concern. This section presents derivations of robustness for component spread and for temperature spread.

In ICs, typically two types of spread occur. Firstly there is intra-die spread, or component mismatch, that results in relatively small *relative* spread between components on the same die. This relative spread is typically lower than 1% and can be neglected with respect to the second type of spread. This second type of spread is a inter-batch spread that results in a significant spread in component values, that strongly correlate per die. This inter-batch spread can amount to 20% shift in the RC-products in our equalizer, whereby all RC products shift in the same direction. Figure 4.9 shows the impact of this spread on the equalizer characteristic.

The gain error due to a correlated shift of the whole equalization curve can be estimated by combining the shift and the slope of the equalization-curve. A

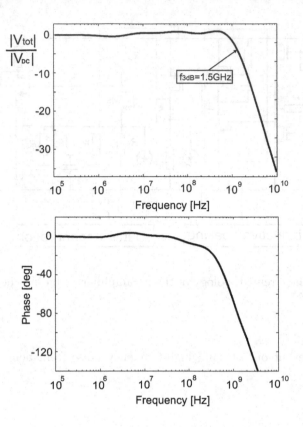

Figure 4.8: The simulated amplitude and phase responses of the photodiode and pre-amplifier after the analog equalizer.

frequency shift by a factor $(1 + \Delta)$ of the whole curve yields a gain error:

$$\frac{\Delta gain}{gain} \approx (1 + \Delta)^{-s} - 1$$

where $s$ is the roll-off of the equalized characteristic. Expressing the frequency change and gain change in dB,

$$\Delta gain[dB] \approx -\Delta[dB] \cdot s \qquad (4.9)$$

As an example, -20% and 20% spread for the total equalization curve yields, at a intrinsic roll-off of -4 dB/decade, a gain spread of only +0.4 dB respectively -0.3 dB in the overall response. Furthermore, it is important to note that the error in the total frequency response of the system is in a small frequency band,

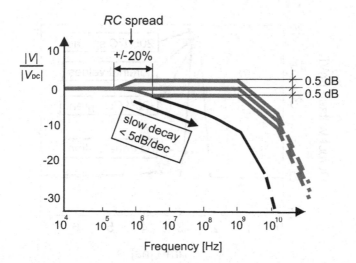

Figure 4.9: An asymptotic approximation of correlated ±20% shift in the RC products in the equalizer on the total system response. The lower curve is the non-equalized response, the higher curves are the nominal equalized response and the +20% and -20% responses.

located around the cut-off frequency of the photodiode. An error in the total equalization characteristic results in ISI only if input frequencies are present in this frequency range. In our system the gain-errors are around 1 MHz, while the bit-rate is around 3 Gb/s: the low gain error due to spread result in only a very small increase in the ISI which can be compensated by a very small increase in optical input power. The main effect of component spread, and the resulting shift in the equalization characteristic, is a changed gain.

These findings are illustrated by the change in the time-response on an (optical) bit (with a square-wave shape and 0.5 ns pulse duration), shown in figure 4.10. The upper three curves in figure 4.10 show the output signal of the optical receiver including equalization, with -20%, 0% and +20% spread in the filter poles with respect to our nominal design. It follows that the effect of this spread is relatively small. As comparison, the lower curve corresponds to the same system, now with a by-passed equalizer.

It can be concluded that the proposed pre-amplifier theoretically is very robust against spread, thanks to the low roll-off in the diode characteristics.

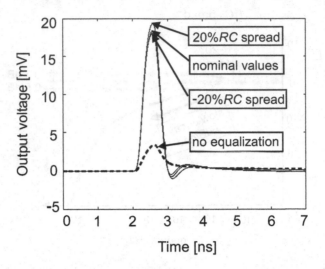

Figure 4.10: Simulated symbol response of the optical detector: for the nominal case, with and without spread and for the non-equalized case

**Robustness on temperature**

In the proposed optical detector, the *intrinsic* response of the photodiode is equalized. This intrinsic response is due to the combined effects of many diffusion currents; this response hence depends on the diffusion constants for carriers. These diffusion constants, in turn, depend mainly on doping levels and on the temperature. The dope-level dependency manifests itself in different temperature-dependencies for the various parts in the photodiode. For the dominant current contribution, the substrate current, the temperature dependency follows directly from the Einstein relation and well known expressions for carrier mobility, e.g. in [18]:

$$\mu_n \propto T^{-2.3\pm0.1} \quad \mu_p \propto T^{-2.2\pm0.1}$$
$$D_n \propto T^{-1.3\pm0.1} \quad D_p \propto T^{-1.2\pm0.1}$$

It follows from these relations that for a temperature range from e.g. 230 K to 370 K the diffusion constant are changed by 38% respectively -30% with respect to that at room temperature. With the earlier findings for spread in equalizer parameters this yields a (deterministic) gain error of up to ±0.6 dB. Concluding: theoretically the system is also inherently robust against temperature variations.

# 4.6 Experimental results

This section discusses the setup and the results of many measurements done on the designed optical receiver system. First the most relevant details of the designed circuit and the measurement setup are discussed. Then a number of measurements that verify the correct behavior of the circuit are given.

## 4.6.1 Circuit details and measurement setup

The design of the optical receiver, the photodiode with pre-amplifier and equalizer,it quite straight-forward. The most relevant details are:

$$\left(\frac{W}{L}\right)_{M1} = \frac{133}{0.18}$$
$$I_{D,M1} = 7mA \qquad g_{m,M1} = 48mS$$
$$C_{in,M1} = 0.5pF \qquad R = 850\Omega$$

Figure 4.11 shows the chip micrograph of the integrated optical detector, including the nwell/p-substrate photodiode and the pre-amplifier with equalizer. As discussed in section 4.3, a minimal nwell-distance finger photodiode with 2 $\mu$m finger size is used as a photodetector. The size of the photodiode is $50 \times 50$ $\mu$m$^2$ yielding a junction capacitance equal to 1.6 pF. The power-supply voltage was

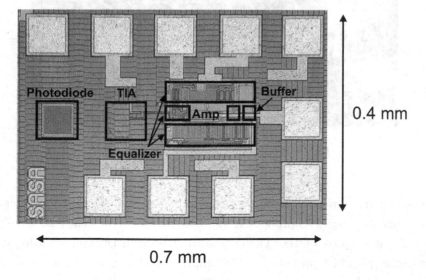

Figure 4.11: Chip micrograph of the optical receiver.

1.8 V. The complete optical detector system consumes approximately 34 mW + 16 mW for the 50 Ω output buffer for evaluation. The total circuit area is $145 \times 305$ µm$^2$. The overall area including bondpads is $0.7 \times 0.4$ mm$^2$.

An 850 nm VCSEL LV1001 from OEPIC company [13] is used as a light source. The light was coupled from the laser to the photodiode using multi-mode fiber with 50 µm core-diameter, with 1 m length. The AC optical power at the fiber's output is deduced by measuring the DC optical power around the operating point of the laser. The optical power is measured using HP 8153A Lightwave Multimeter. The shape of the output signal and the AC optical power were also measured using the reference photoreceiver PT1003 from OEPIC company, which consists of a PIN photodetector, integrated with an InGaPHBT transimpedance amplifier (TIA). The maximum operating data-rate of this reference photoreceiver is 10 Gb/s.

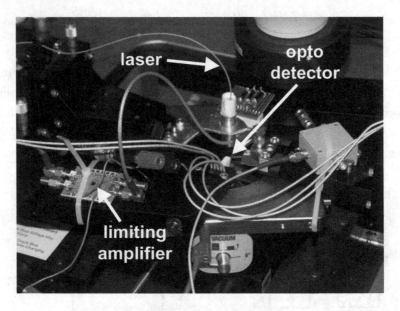

Figure 4.12: The measurements set-up.

The on-chip measurements were done using probe-station. The output voltage is measurements using GSG probe ACP40. The insertion loss of the coaxial cables was calibrated up to 4 GHz. The chip is supplied with DC voltage using Eye-pass probe [16] which provide a stable-supply in the frequency range of interest (the specified frequency range is up to 20 GHz). The DC *current* supply was set using coaxial cables and GS pico-probes.

The laser was modulated with the pseudorandom bitstream of $2^{31}$-1 from the Anritsu MP1632C digital data analyzer. Since the swing of the signal at the pre-amplifier's output was not large enough for proper BER measurements, a limiting amplifier was placed after the pre-amplifier as shown in figure 4.13. The limiting amplifier is an L1001 fabricated by OEPIC company.

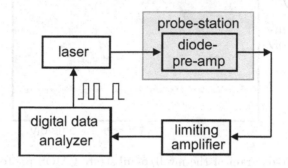

Figure 4.13: Block schematic of the measurements set-up.

## 4.6.2 Optical receiver performance without equalizer

Figure 4.14 shows the eye diagram of the integrated pre-amplifier *without* equalization. For this system, theoretically the maximal speed for BER<$10^{-11}$ is 10 Mb/s. An eye-diagram for 50 Mb/s input with BER= $10^{-7}$ is measured since that is a minimum speed of the used digital data analyzer. The input light power is $25\mu$W peak-to-peak (-19 dBm) optical power. The measurements on the system without equalizer (by disabling the 4 zeros in the circuitry that otherwise take care of the equalization) were done to clearly see the effect of the equalization. All other measurements are done using receivers with enabled equalization.

## 4.6.3 Optical receiver performance with equalizer

The eye-diagram shown in figure 4.15 shows the performance of the optical receiver system with equalization. For the results in figure 4.15, the input light power (AC) is again 25 µW peak-to-peak and the data rate is 3 Gb/s; the achieved BER<$10^{-11}$. This BER figure is one order of magnitude larger than the theoretical value, which is due to excess noise in the circuit supply and in the limiting amplifier.

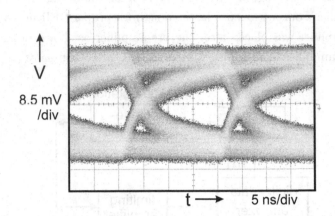

Figure 4.14: Eye-diagram of the nwell/p-substrate CMOS photodiode without equalizer 50 Mb/s, BER=$10^{-7}$.

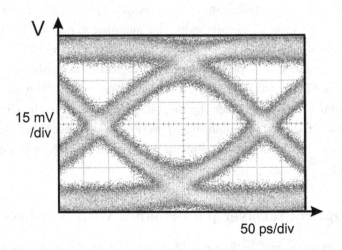

Figure 4.15: Eye-diagram of the nwell/p-substrate CMOS photodiode with an analog equalizer 3Gb/s, BER=$10^{-11}$.

Note that the usage of the analog equalizer resulted in orders of magnitude increase in data rate at orders of magnitude lower BER.

The leftmost curve in figure 4.16 shows the measured BER as a function of the light input power for 3 Gb/s. The sensitivity at the BER of $10^{-11}$ is around -19 dBm, which is 2 dBm better than the one defined in the Gigabit Ethernet standard for short-haul optical communications [12]. The rightmost curve is discussed in the section dealing with robustness against photodiode-spread.

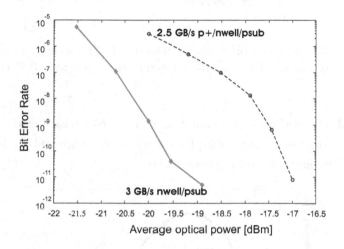

Figure 4.16: Bit error rate as a function of the input optical power

### 4.6.4 Robustness of the pre-amplifier: component spread

The theoretical high robustness of the pre-amplifier circuit on spread is confirmed with the measurements on the circuit shown in figure 4.11. During the chip-layout design, all $RC$ filter components are placed as a number of smaller components (fingers) connected using the highest metal layer. Removing some of this metal-layer connections, the $RC$ values can be changed. In the experiment, we changed the values for $\pm 20\%$ after the fabrication using the Focused Ion Beam (FIB). The eye-diagrams are measured at the output of the pre-amplifier for 3 Gb/s data-rate, and the results are shown in figure 4.17.

The impact of the spread on the performance of the optical detector system is more easily seen in Figure 4.18. The curve in that figure shows the simulated eye-amplitude at the output of the detector, as a function of spread in the equalizer's zeros. Our nominal design is indicated by the vertical dotted line;

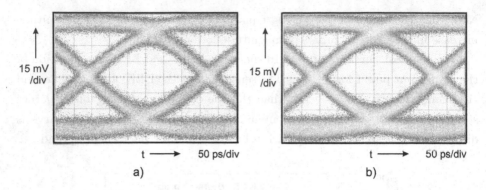

Figure 4.17: Eye diagrams on 3 Gb/s data-rate of the pre-amplifiers including a) $+20\%RC$ spread and b) $-20\%RC$ spread in the equalizer. BER is $10^{-10}$.

note that this design is non-optimum which is due to a design error. The dots in the figure are measurement data for our nominal design and for $+20\%$ and $-20\%$ variation on the zeros in the equalizer.

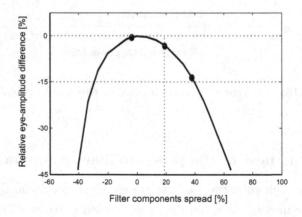

Figure 4.18: Simulated relative eye-amplitude change at the equalizer's output as a function of the spread in $RC$, and some measurement results (dots).

## 4.6.5   Robustness of the pre-amplifier: diode spread

To measure the impact of photodiode spread, the optical detector circuit was also implemented using a different photodiode, with the same pre-amplifier circuit. This section presents the pre-amplifier integrated with p+/nwell/p-substrate photodiode (double photodiode). The same pre-amplifier and analog equalizer used with nwell/p-substrate diode (shown in figure 4.7) are used here. The filter parameters in the equalizer are therefore not optimized for the double photodiode characteristics shown in figure 3.26. In this manner, the robustness for (very large) spread in photodiode characteristics is measured.

The measured eye-diagram for for the optical receiver with the p+/nwell/p-substrate photodiode is presented in figure 4.19. The achieved data-rate is 2.5 Gb/s with 38.5 µW optical power. With the expense of approximately 2 dB higher input optical power with respect to using the optimal photodiode, but without optimizing the equalizer, a very high data-rate is achieved. By optimizing the HF filter parameters, the simulated data-rate of the system is also 3 Gb/s for the previously used optical power of 25 µW.

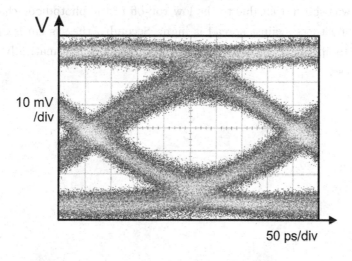

Figure 4.19: 2.5 Gb/s eye-diagram of the p+/nwell/p-substrate CMOS photodiodes with same analog equalizer used for nwell/p-substrate diode, BER=$10^{-11}$.

**Temperature measurements**

All the previous measurement results were obtained at room temperature. The sensitivity to the temperature of the optical detector was determined using BER and eye-amplitude measurements for a number of temperatures. For a change in temperature of 25 K the measured photosensitivity at 3 Gb/s data rate decreases by only 0.3 dB with respect to that at room temperature. At 75 K temperature increase the decrease in sensitivity amounts to 1.7 dB. These results confirm that the optical detector system is fairly robust against temperature changes.

Because the temperature deterministically affects the photodiode response, it could be minimized by a simple feed-forward control network.

## 4.7   Conclusions

The proposed optical detector architecture with an analog equalizer can be used to increase the bit rate by several orders of magnitude. Compared to state of the art CMOS detectors such as in [1] the bit rate increment is about 4.5 for $\lambda = 850$ nm, without reducing the photo-responsivity. A 3 Gb/s data-rate is achieved with 25 µW peak-to-peak light input power and BER<$10^{-11}$.

The high-speed optical detector with an analog equalizer is very robust. Firstly, it was shown that due to the low roll-off of the photodiode characteristics, the robustness against spread is high. Secondly, the system is inherently robust for temperature variations while it is possible to automatically compensate for these.

# Bibliography

[1] D. Coppée, H. J. Stiens, R. A. Vounckx, M. Kuijk: "Calculation of the current response of the spatially modulated light CMOS detectors", *IEEE Transactions Electron Devices*, vol. 48, No. 9, 2001, pp. 1892-1902.

[2] H. Lin, R.C. Chang, H. Chih-Hao, L. Hongchin: "A flexible design of a decision feedback equalizer and a novel CCK technique for wireless LAN systems Circuits and Systems", Proceedings of the 2003 International Symposium on Circuits and Systems (ISCAS'03), Volume: 2, 25-28 May 2003, pp.II 153-II 156.

[3] D. Inami, Y. Kuraishi, S. Fushimi, Y. Takahashi, Y. Nukada, S. Kameyama, A. Shiratori: "An adaptive line equalizer LSI for ISDN subscriber loops", *IEEE Journal of Solid-State Circuits*, Volume 23, Issue 3, June 1988, pp.657-663.

[4] W.L. Abbott, H.C. Nguyen, B.N. Kuo, K.M. Ovens, Y. Wong, J. Casasanta: "A digital chip with adaptive equalizer for PRML detection in hard-disk drives", *IEEE International Solid-State Circuits Conference*, Digest of Technical Papers, 16-18 Feb. 1994, pp.284-285.

[5] A. Bishop, I. Chan, S. Aronson, P. Moran, K. Hen, R. Cheng, L.K. Fitzpatrick, J. Stander, R. Chik, K. Kshonze, M. Aliahmad, J. Ngai, H. He, E. daVeiga, P. Bolte, C. Krasuk, B. Cerqua, R. Brown, P. Ziperovich, K. Fisher:

"A 300 Mb/s BiCMOS disk drive channel with adaptive analog equalizer", *IEEE International Solid-State Circuits Conference*, Digest of Technical Papers, 15-17 Feb. 1999, pp.46-49.

[6] K. Azadet, E. Haratsch, H. Kim, F. Saibi, J. Saunders, M. Shaffer, L. Song, and M. Yu: "Equalization and FEC techniques for optical transceivers", *IEEE Journal of Solid-State Circuits*, vol. 37, March 2002, pp. 317-327.

[7] S.Radovanovic, A.J.Annema, and B.Nauta "3 Gb/s monolithically integrated photodiode and pre-amplifier in standard 0.18um", in Dig. Tech. Papers, ISSCC 2004, pp. 472-473

[8] S. Radovanovic, A.J. Annema and B. Nauta, "A 3-Gb/s Optical Detector in Standard CMOS for 850-nm Optical Communications", *IEEE Journal of Solid-State Circuits*, vol. 40, August 2005

[9] W. Etten and J.vd Plaats: *"Fundamentals of Optical Fiber Communications"*, Prentice-Hall, 1991.

[10] B. Razavi: *"Design of Analog CMOS Integrated Circuit"*, McGraw-Hill Higher Education, 2001.

[11] S. Alexander: *"Optical communication receiver design"*, SPIE Optical engineering press, 1997.

[12] IEEE 10 Gigabit Ethernet Standard 802.3ae.

[13] http://www.oepic.com/hm021206/Products.asp

[14] C. H. Lu, W. Z. Chen: "Bandwidth enhancement techniques for transimpedance amplifier in CMOS technology", *ESSCIRC 2001*, 18-20 September 2001, Villach, Austria, pp.192-195.

[15] P.Amini and O.Shoaei: "A low-power gigabit Ethernet analog equalizer", *ISCAS 2001*, pp 176-179.

[16] http://www.cascademicrotech.com/index.cfm/fuseaction/pg.view/pID/124

[17] T. van der Meer: *"Design of a PDIC for CD/DVD-system in standard CMOS technology"*, Master Thesis, June 2004.

[18] R.F. Pierret, "Semiconductor Fundamentals", Addison-Wesley: 1989

# CHAPTER 5

# Bulk CMOS photodiodes for $\lambda = 400$ nm

*The photodiode bandwidth is a strong function of the wavelength. The previous two chapters assumed $\lambda$=850 nm. This chapter presents both time domain and frequency domain analyses of monosilicon photodiodes in a standard 0.18 μm CMOS technology, for $\lambda = 400$ nm.*

*For monosilicon diodes, the maximum calculated intrinsic -3 dB bandwidth is up to 6 GHz at $\lambda = 400$ nm; this corresponds to a cut-off frequency of about 4 GHz. The photodiodes designed in twin-well technology have smaller bandwidth because of the limited size of the vertical depletion region. Measurements on p+/nwell/p-substrate photodiode designed in 0.18 μm CMOS, showed that the total diode bandwidth is 1.7 GHz, which was limited by the electrical diode bandwidth in our measurements.*

## 5.1    Introduction

Section 2.9 showed that the lower the wavelength of the input signal, the higher
the light absorption coefficient $\alpha$. For the lower and upper limit of the CMOS
sensitivity range $\lambda \in [400,850]$ nm, the difference in light penetration depth is
almost 70 times. At the lower boundary ($\lambda = 400$ nm), the 1/e-absorption
depth is only about 0.2 μm. In both previous and modern CMOS processes
(up to 0.13 μm technology), this depth is certainly less than or equal to the
shallowest junction available (n+ or p+). Hence, light is absorbed very close
to the diode surface. As a result, the overall photocurrent is determined by the
(fast) diffusion inside n+/p+/nwell regions and the drift photocurrent generated
in the vertical depletion regions as shown in figure 5.1.

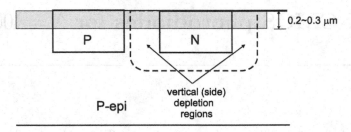

Figure 5.1: Light absorbtion in silicon photodiode for $\lambda = 400$ nm.

The responsivity of a CMOS photodiode and hence the photocurrent, is low
for $\lambda = 400$ nm: the energy of the incoming photons is $h\nu$ and for the same input
optical power $P_{in}$, the number of photons $P_{in}/h\nu$ is minimal. This also follows
from figure 2.11. As a result of this relatively low number of photons, the photocurrent is relatively low. In addition, the surface recombination process is now
important: due to the low light-penetration depth, the surface recombination
of the carriers is significant. At 400 nm the maximum photodiode responsivity
is about 0.23 A/W.

Due to the very low penetration depth of the light, the lateral photodiode
structure[1] becomes important in the overall photodiode response [2, 3]. There
are two main advantages of using lateral structures:

- for diodes for which large intrinsic (depletion) regions are the dominant
  in the structure, most of the carriers are generated in this region. As a
  result, the total photodiode bandwidth can be tens of GHz. For diodes

---

[1]The structure along the $y$-axis in figure 3.2.

without a significant intrinsic layer, carriers are generated in the n-region and the p-region close to the diode surface. Depending on the depth of the n and p-regions, the diffusion of these carriers can be fast, which results in a large diode bandwidth; this will be shown in the following part of this chapter.

- surface recombination is less dominant for carriers generated incidently in the vertical (side) depletion region.

The lateral photodiodes in standard CMOS technology with maximal vertical (side) junctions can be designed by making the nwell separate to the pwell; this provides large depletion region between the wells by exploiting p-epi layer in between. Both the frequency and the time response of the twin-well photodiodes are analyzed in sections 5.2, 5.3 and 5.5. For comparison, a separate-well nwell/p-substrate photodiode is analyzed in section 5.4.

# 5.2 Finger nwell/p-substrate diode in *adjoined-well* technology

The nwell/p-substrate diode in *twin-well* technology is shown in figure 5.2. In this chapter we present the photodiode frequency response and time response on a Dirac light pulse for $\lambda = 400$ nm. Both responses are calculated following the procedure explained in chapter 3. The absorption depth of light is smaller than the junction depths so the effect of the substrate current component is negligibly small. To simplify the analyzes, the nwell diffusion current is analyzed in detail, while the (complementary) pwell diffusion is approximated by a scaled version of its nwell complement.

A few ps after an incident light pulse, most of the excess carriers are generated close to the photodiode surface. Charge diffusion results from charge density gradients see e.g. chapter 3. At short wavelengths, the vertical gradient of excess holes in the nwell typically is lower than the lateral gradient at narrow nwells and $\lambda$=400 nm. For wide nwells, the effective lateral gradient is low and the vertical gradient is dominant. For narrow nwells, the lateral dimension is dominant for the diode speed at $\lambda$=400 nm, while for (slower) wide nwells the vertical dimension is dominant. An illustration for the diffusion in a narrow nwell is given in figure 5.3.

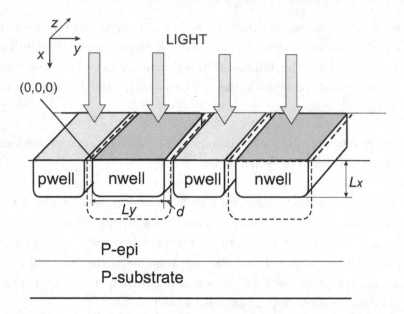

Figure 5.2: Finger nwell/p photodiode structure with *low resistance* substrate and adjoined-wells in standard CMOS technology.

The total photocurrent is the sum of the diffusion current (3.8), and the drift current (3.17). The nwell/p-substrate photodiode frequency response is shown in figure 5.4. The response is normalized with the DC photocurrent shown at 400 nm. Figure 5.4 shows the importance of the nwell width on the diode intrinsic bandwidth. This -3 dB bandwidth is about 700 MHz for minimum nwell width[2] (2µm) and about 300 MHz for a wide nwell, 10 µm. Therefore, *for maximum intrinsic bandwidth, the diode nwell size should be minimal.* Note that the roll-off of the responses in this chapter are much higher than the roll-offs in previous chapters. This is due to the fact that the wavelength is low: for low wavelengths the different current contribution do not nicely sum to obtain an overall low roll-off. The above mentioned -3 dB bandwidths correspond to a cut-off frequency that is about a factor $\sqrt{3}$ lower (assuming a roll-off of 10dB/decade).

---

[2]In standard CMOS the minimum nwell width is typically about twice its depth.

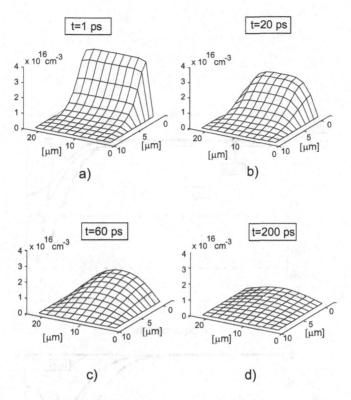

Figure 5.3: The calculated hole diffusion profile inside the nwell for a minimum nwell width, 2 μm, under incident light pulse ($\lambda = 400$ nm, 10 ps pulse-width). This profile is calculated after 1 ps, 20 ps, 60 ps and 200 ps.

## 5.3  Finger n+/nwell/p-substrate diode

In standard CMOS technology, it is possible to place a shallow n layer (n+), at the top of the nwell region as shown in figure 5.5. This section discusses the impact of such an n+ layer inside the nwell on the total photocurrent response for $\lambda = 400$ nm.

The depth of the n+ layer in standard 0.18 μm CMOS technology is larger than the 1/e-absorbtion depth at $\lambda = 400$ nm. The frequency response of the n+/nwell diffusion current is calculated using (3.10).

To solve the system of equations, two boundary conditions between the two n-layers are used, plus the one boundary at the n+ top and the boundary at nwell bottom, as shown in figure 5.6. These conditions are related to both the current density and the minority carrier concentration. Due to the continuity

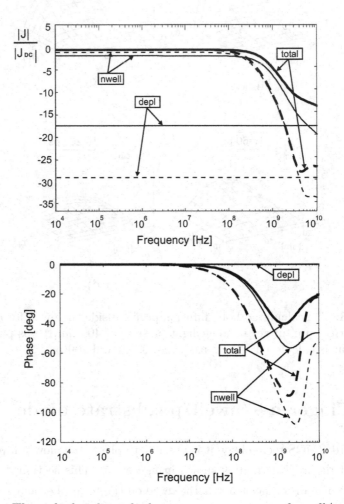

Figure 5.4: The calculated total photocurrent response of nwell/p-substrate photodiodes in a *twin-well* technology for the minimum nwell width (solid-line) and nwell width much larger than its depth 10μm (dashed-line) for $\lambda = 400$ nm.

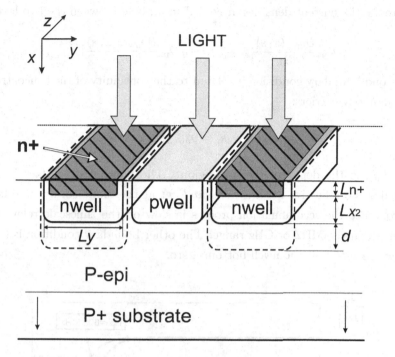

Figure 5.5: Finger nwell/p-substrate photodiode with n+ layer at the top of the n-well region.

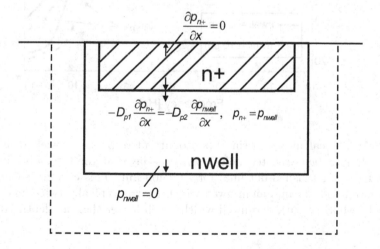

Figure 5.6: The boundary conditions for the calculations of nwell/n+ diffusion current.

of currents, the current densities are equal in a plane between the two layers:

$$-qD_{p1}\frac{\partial p_{n+}(x,s)}{\partial x}\Big|_{x=L_{n+}} = -qD_{p2}\frac{\partial p_{nwell}(x,s)}{\partial x}\Big|_{x=L_{n+}} \tag{5.1}$$

The second boundary condition is related to the continuity of the concentration of the minority carriers:

$$p_{n+}(L_{n+},s) = p_{nwell}(L_{n+},s) \tag{5.2}$$

where $L_{n+}$ is the depth of the n+ region. The photodiode surface is again assumed to be reflective i.e. the *hole gradient* at the diode surface is taken to be zero, since the recombination process is slow on the timescale relevant for frequencies in the MHZ or GHz range. The other boundary condition is for the *hole concentration* at the nwell bottom: zero.

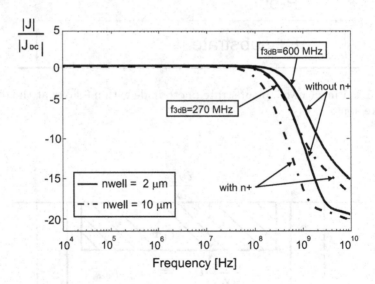

Figure 5.7: The calculated intrinsic photocurrent response of nwell/p-substrate photodiode in a *twin-well* technology *with* n+ layer at the top (solid line) and *without* n+ layer (dashed-dot line) for $\lambda = 400$ nm. This current is calculated for two nwell widths: minimum nwell width that is typically twice its depth in standard CMOS (2 µm), and nwell width much larger than its depth (10 µm).

The total n+/nwell p-substrate frequency response for $\lambda = 400$ nm is shown in figure 5.7. The maximum -3 dB bandwidth of this photodiode is about 270 MHz for 2 µm nwell width and 180 MHz for 10 µm wide nwells. These bandwidth are twice as low as the corresponding bandwidths of the photodiode without n+ layer because the diffusion constant $D_{p1}$ is twice as low for the n+ region than $D_{p2}$ for the nwell region due to the higher majority carrier concentration. Thus, a highly doped n+ region at the top of the nwell (nwell/p-substrate diode) *decreases* maximum intrinsic bandwidth for more than *two times*. Note that the roll-off of these diodes at 400 nm light are big: between -10 dB/decade and -20 dB/decade.

For a photodiode area of $50 \times 50$ $\mu$m$^2$, corresponding to the core-diameter of the multimode fiber, the photodiode capacitance is given in Table 3.5. It ranges from 0.6-1.6 pF for 10 µm and 2 µm nwell size, respectively. For the input resistance of the subsequent transimpedance amplifier lower than 130 $\Omega$, the extrinsic photodiode bandwidth is larger than the intrinsic diode bandwidth. For larger transimpedances, the nwell size i.e. photodiode capacitance does influence the total photodiode bandwidth. The lower the nwell width, the lower the electrical bandwidth.

## 5.3.1   Time domain measurements

The calculated total photodiode bandwidth is confirmed by measurements on a minimum width photodiode in a 0.18 $\mu$m CMOS process. The diode layout is given in figure 5.8. On the transmitter side, a picosecond blue-light laser with $\lambda = 400$ nm was used. The light was focused into a multimode fiber using a system of lenses, as shown in figure 5.9. The pulse width of the picosecond laser is 200 ps and the power is 1 mW. The output voltage of the n+/nwell p-substrate photodiode is measured using RF pico-probe and coaxial cable which was terminated with 50$\Omega$ of the oscilloscope. This voltage is shown in figure 5.10.

Because the roll-off of the photodiodes at 400nm light is relatively large, even approaching -20 dB/decade, the photodiode bandwidth can be estimated using well known formulas that hold for first order systems. For first order systems e.g.

$$f_{3dB} \approx \frac{ln(9)}{\pi(\tau_r + \tau_f)} \quad [2]$$

$$f_{3dB} \approx \frac{ln(9)}{2\pi\tau_f}$$

$$f_{3dB} \approx \frac{1}{2\pi\tau_{37\%}}$$

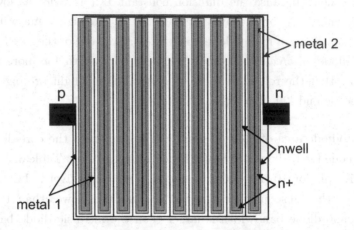

Figure 5.8: A nwell/n+/p-substrate photodiode with 2 µm nwell width.

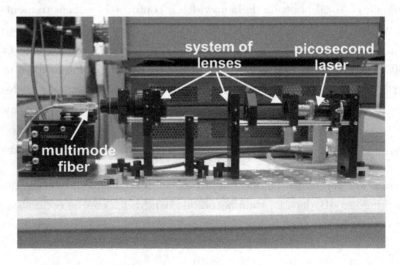

Figure 5.9: Focusing the light from the picosecond blue laser into the multimode fiber by using system of lenses.

In these equations, $\tau_r$ is the rise time, $\tau_f$ is the fall time and $\tau_{37\%}$ is the time duration to fall to 37% of the starting value. It follows from the measurements that the bandwidth is about 230 MHz, which complies to the calculation result shown in figure 5.7.

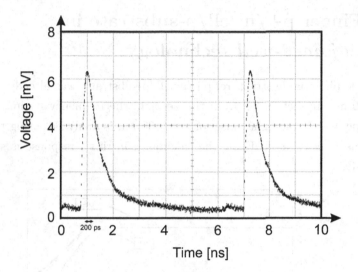

Figure 5.10: Transient response of the nwell/n+ p-substrate photodiode on 200 ps input light pulse($\lambda$ = 400 nm) with a 50 $\Omega$ load resistor; 2 $\mu$m wide fingers.

## 5.4   Finger n+/p-substrate photodiode in *separate-well* technology

The frequency analysis of n+/p-substrate photodiode is similar to the previously analyzed photodiode. However, the main difference is in the size of the vertical depletion regions between n+ fingers and p-epi region; here it is larger, which results in a faster total intrinsic response. The calculated intrinsic -3 dB bandwidth for a n+/p-subs photodiode in 0.18 $\mu$m CMOS is 6 GHz. In [2], a 9-finger n+/p-substrate photodiode is presented for high-speed data rate. The photodiode was designed in standard 1-µm CMOS technology. The doping concentration of the epitaxial layer in the used CMOS technology is very low ($< 10^{15}$ cm$^{-3}$). On the other hand, the doping concentration of the shallow n+ region is very high ( $10^{20}$ cm$^{-3}$) resulting in a large depletion region. The measured bandwidth for 400 nm wavelength was 470 MHz and was limited by the electrical photodiode bandwidth (the resistance subsequent to the photodiode is 1 kΩ).

## 5.5   Finger p+/nwell/p-substrate in *adjoined-well* technology

The double photodiode structure, p+/nwell/p-substrate, was already analyzed for λ=850 nm, in section 3.2.4. In this section, the frequency response of this photodiode on a Dirac light pulse for $\lambda = 400$ nm is analyzed, using the same analyses as in chapter 3. First, the calculated intrinsic response is shown in figure 5.11.

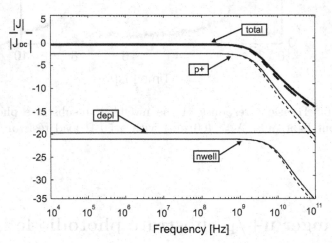

Figure 5.11:   The calculated total photocurrent response of p+/nwell/p-substrate photodiode in a *adjoined-well* technology with 2 μm nwell size (solid-line) and 10 μm nwell size (dashed-line) for $\lambda = 400$ nm.

Most of the carriers are again generated close to the photodiode surface inside the p+ region and the vertical (side) depletion regions. Therefore, the diffusion process is fast and the calculated bandwidth using equation (3.8) is 2 GHz. The total -3 dB bandwidth including the drift current in vertical junctions is 2.8 GHz.

*Electrical bandwidth*

For a photodiode area of $50 \times 50$ $\mu m^2$, the photodiode capacitance is given in Table 3.5. It ranges from 2.2 pF-3.6 pF for 10 µm and 2 µm p+/nwell size, respectively. For an input resistance of the subsequent transimpedance amplifier (TIA) lower than 15 Ω, the extrinsic photodiode bandwidth is larger than intrinsic diode bandwidth.

## 5.5.1 Time domain measurements

The response of the single p+/nwell/p-substrate diode is measured in the time domain. The top view of this photodiode is shown in figure 5.12. A picosecond blue-light laser with $\lambda = 400$ nm was used as a signal source. The pulse width of the picosecond laser is 200 ps and the maximum optical power is 1 mW. The output signal is measured using RF pico-probe and the coaxial cable terminated with 50 Ω load of oscilloscope. The bondpads for n-contact and p-contact and reversed in comparison with the previous diode. Therefore, the output voltage presented in figure 5.13 is also inverted.

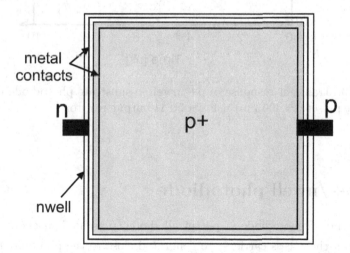

Figure 5.12: Top view of the single p+/nwell/p-substrate photodiode.

Using the output-voltage time response, the approximated *overall* -3 dB bandwidth of the single p+/nwell/p-substrate photodiode (5.3) is 1.4 GHz. According to the calculations, the *electrical* diode bandwidth (first order) including diode capacitance and the output impedance ($R_{out}$=50 $\Omega$, $C_{in}$=1.7 pF) is 1.7 GHz, while the *intrinsic* diode bandwidth is 2.8 GHz (see figure 5.11). Therefore, the overall bandwidth should be about 1.7 GHz which comply with measurements when the bondpad capacitances ($\sim$100 fF) are also taken into account in the electrical bandwidth.

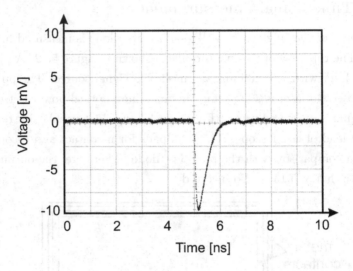

Figure 5.13: Transient response of p+/nwell/p-substrate photodiode on 200 ps input light pulse($\lambda = 400$ nm) using a 50 $\Omega$ output resistor.

## 5.6   p+/nwell photodiode

Another type of photodiode is p+/nwell diode, presented in chapter 3. For $\lambda = 400$ nm, the excess carriers are generated close to the photodiode surface. As a result, the diode substrate will not significantly contribute to the overall photocurrent. By disconnecting the substrate, the photodiode capacitance decreases, which increases the electrical photodiode bandwidth at no responsivity penalty.

The width of the p+/nwell fingers does not influence the intrinsic photodiode bandwidth because of the shallow depth of the p+ region, as was described in section 5.5. On the other side, the larger the size of the p+/nwell fingers the lower the diode capacitance. It follows that the maximum intrinsic and extrinsic bandwidth are obtained using single wide nwell finger.

The capacitance of the single p+/nwell photodiode with dimensions 50 µm × 50 µm is 1.7 pF for 0.18 µm CMOS. For an input resistance of the subsequent TIA lower than 30 Ω the electrical diode bandwidth is larger than the intrinsic diode bandwidth.

## 5.7 Conclusion

For the lower boundary of the CMOS wavelength-sensitivity range $\lambda = 400$ nm, the -3 dB bandwidth of the photodiodes in 0.18 µm CMOS technology is in the range from 170 MHz to 6 GHz; the roll-off at 400 nm is much higher than the roll-off at longer wavelengths and easily amounts to -10 dB/decade.

The maximum intrinsic bandwidth of 6 GHz, is achieved with nwell/p-substrate photodiode designed in separate-well technology because of the maximum depletion region area. Using the adjoined-well technology, the maximum calculated intrinsic bandwidth is about 3 GHz and is achieved using a single p+/nwell photodiode. The influence of the nwell/p+ width is negligible on the intrinsic bandwidth, because the bandwidth is determined by the shortest distance for the diffusion process: the (small) p+ depth. The number of nwell/p+ fingers however, does influence the overall bandwidth: the higher the number of fingers the lower the nwell/p+ width and the higher the photodiode capacitance.

For the nwell/p-substrate photodiode in the adjoined-well technology, the nwell width is very important in the diode intrinsic bandwidth. The highest calculated bandwidth is 700 MHz achieved with a minimal nwell-width (2 µm). Larger nwell widths decrease the diode intrinsic bandwidth by almost a factor two. In addition, the n+ layer, which may exist at the top of the nwell, decreases the diode bandwidth further by a factor two. This is because the high majority carrier concentration inside n+ decreases the minority carrier diffusion constant and thus, the bandwidth. Maximum bandwidth of nwell/p-substrate photodiode is obtained with minimum width of the nwell region and without an n+ layer at the nwell-top.

# Bibliography

[1] Wei Jean Liu, Oscal T.-C. Chen, Li-Kuo Dai and Far-Wen Jih Chung Cheng: "A CMOS Photodiode Model", *2001 IEEE International Workshop on Behavioral Modeling and Simulation*, Santa Rosa, California, October 10-12, 2001.

[2] H. Zimmermann, H. Dietrich A. Ghazi, P. Seegebrecht: "Fast CMOS Integrated Finger Photodiodes for a Wide Spectral Range", *ESSDERC 2002*, pp. 435-438, 24-25 September, Italy

[3] S. M. Sze: *"Physics of semiconductor devices"*, New York: Wiley Interscience, 2-nd edition, p. 81, 1981.

[4] B. Razavi: *"Design of Analog CMOS Integrated Circuits"*, McGraw-Hill, 2001.

# CHAPTER 6

# Polysilicon photodiode

*This chapter presents a lateral polysilicon photodiode that has an intrinsic bandwidth far in the GHz range. The electrical bandwidth is also high due to a very small parasitic capacitance (<0.1 pF). For $\lambda = 400$ nm, the achieved quantum efficiency is however only 2.5% due to the very small light sensitive diode volume. The diode active area is limited by a narrow depletion region while the small depth is limited by the technology.*

## 6.1 High-speed lateral polydiode

In nowadays CMOS processes, a polycrystalline silicon (polysilicon) layer is available above the silicon-oxide; this is typically used as a gate terminal for both NMOST and PMOST. The doping concentration of this polysilicon layer is high $(1 \cdot 10^{20} \text{cm}^{-3})$ with the doping charge corresponding to the type of the MOS transistor. Using these two opposite types of polysilicon we made a polysilicon photodiode [1], see figure 6.1.

The main advantage of the polysilicon photodiode in comparison with the monosilicon one is that there are no slow-diffusive carriers coming from the substrate, for all wavelengths of interest.

123

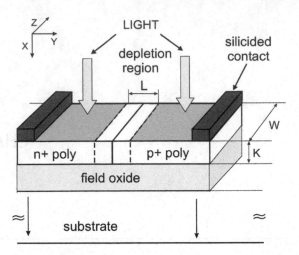

Figure 6.1: Lateral polysilicon photodiode in CMOS technology.

The total diode response is the sum of two responses: fast drift current inside the depletion region and fast diffusion current inside n+ and p+. The latter is due to the high doping concentration inside n+ and p+. The lifetime of the excess carriers is low (ps range [2]), the diffusion lengths $L_{n,p}$ are short (around 300 nm [2]) and only carriers generated sufficiently close to the junctions are collected as photocurrent; the rest of the excess carriers is recombined. The main differences of *polysilicon* in comparison with *monosilicon* photodiode concerning photo-responsivity and data-rate are:

- the absorption coefficient $\alpha$ is four times the absorption coefficient of monosilicon photodiodes [2, 3], (for the same material depth, the responsivity of polysilicon diode is higher for all wavelengths $\lambda \in [400, 850]$ nm)

- the electron mobility $\mu_n$ is approximately four times lower than the mobility of monosilicon photodiode [4], which can limit the bandwidth of the polysilicon photodiodes.

The depth of the poly layer in standard CMOS technology is typically less than 600 nm. Therefore, the polysilicon layer is sensitive mainly for short wavelengths ($\lambda < 600$ nm).

The first lateral pn junction in polysilicon found in literature was designed and investigated by J. Manoliu in 1972 [5]. The dopant concentrations on both sides are very high, about $2 - 5 \times 10^{15} \text{cm}^{-2}$. Compared to the p-n junctions in a single-crystal Si, polysilicon diodes carry much higher current densities. For many years after, the knowledge on polysilicon diodes' behavior was maturing and in 1994 [4], a thorough theoretical and numerical analyzes on this diode is presented while the obtained results showed good agreement with measurements. High leakage current in polydiodes was explained with the field enhanced effect, where a large number of carriers typically trapped in the grain boundaries, are released due to a high electric field. A detailed analysis can be found in [4].

According to literature, a PIN polysilicon photodiode was first introduced as a high-speed photodetector in 1994 [6]. The time-response measurements using the high power pulse-laser with $\lambda = 514$ nm, showed that the -3 dB frequency of that polysilicon photodiode was 5 GHz. The doping concentrations of n+ and p+ poly regions were very high: $3.3 \cdot 10^{20} \text{cm}^{-3}$.

In 1997, the PIN polysilicon resonant-cavity photodiode with silicon-oxide Bragg reflectors was introduced with a speed in the GHz range [7]. Doping concentrations of n+ poly is $2 \cdot 10^{20}$ cm$^{-3}$ and p+ poly $4 \cdot 10^{19}$cm$^{-3}$. The absorption thickness of the polysilicon was 500 nm which is 8 times higher in comparison to the previously reported poly-diodes. This also implies that this photodiode is suitable for wavelengths in the range between $\lambda \in [400, 600]$ nm. For 600 nm, the maximal amount of the absorbed light is about 50%. In [7], a quantum efficiency of 40% is reported for input wavelength light $\lambda = 640$ nm. The responsivity measurements as well as the high frequency measurements for $\lambda = 400$ nm were not presented in the paper.

In this chapter we present lateral polysilicon photodiode in standard CMOS technology. The main difference in comparison with the diodes in [6] and [7] are:

- there is no intrinsic (low doped or undoped poly) layer between n+ and p+ regions; as a result, the light sensitive area is smaller.

- the depth of the poly-diode is 0.2 μm i.e. smaller than reported ones;

- designed in standard CMOS technology, this poly-diode can be easily integrated with the rest of the electronic circuitry. This is very suitable for low-cost, high-speed optical detector design. Moreover, an array of detectors can be easily designed which increases the overall data-rate for the cost of minimal chip area (simple embedding). This is valid for all silicon photodiodes.

Figure 6.2 shows the measured I-V characteristic of the polysilicon photodiode
without light. The large leakage current is due to grain-boundary trap-assisted
band-to-band tunnelling and field-enhanced emission [4].

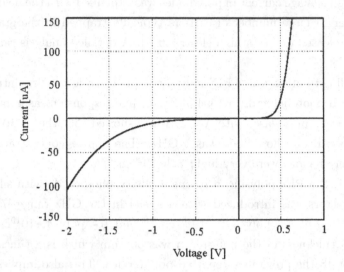

Figure 6.2: Measured DC current (without light) of "jagged" polysilicon photodiode in standard CMOS technology.  The lateral diode dimensions are
$45 \times 45 \ \mu$m.

During the chip processing the masks for the $n+$ and $p+$ layers shown in
figure 6.1 are never perfectly aligned, and dopes tends to diffuse sideways. This
influences the size of the effective width of the polydiode's depletion region.
However, measurements on a number of devices on the same wafer showed that
the effects of misalignment and lateral diffusion as seen as spread in sensitivity
were negligibly small.

The carrier lifetime in polysilicon diode depends on the recombination rate
of carriers and it is proportional to the concentration of recombination centers
[5] and inversely proportional to the grain size of polysilicon. In 0.18 $\mu$m CMOS
technology, the grain size is about 50-60 nm [8], which causes the carrier lifetime
to be very short, about $\tau_{n,p} = 50$ ps [2]. Since in this case the diffusion speed
of carriers is mainly determined by their lifetime, the diffusion bandwidth will
be far in the GHz range ($f \sim 1/\tau_{n,p}$).

## 6.1.1 Pulse response of the poly photodiode

The major speed limitation in all monosilicon photodiodes lies in the very slow diffusion of excess carriers generated deep into the substrate when using long-wavelength light. This section analyzes the intrinsic processes in the poly photo-diode including the drift and the diffusion of carriers inside the depletion region as well as the diffusion of carriers outside this region. The latter is not negligible for narrow poly photodiodes without an intrinsic region.

The current response of polysilicon detector is mainly determined by the minority carrier lifetimes $\tau_n$ and $\tau_p$, saturation drift velocities $v_s$ and diffusion of minority carriers inside depletion region; the last one is important if the width of the depletion region is larger than excess carrier diffusion lengths [6]. If $n(x,t)$ is the excess electron concentration and $p(x,t)$ is the excess hole concentration, the transport of these carriers *inside* the junction can be described with drift-diffusion equations as [9, 6]:

$$\frac{\partial n(x,t)}{\partial t} = D_n \frac{\partial^2 n(x,t)}{\partial x^2} \pm v_n \frac{\partial n(x,t)}{\partial x} - \frac{n(x,t)}{\tau_n} + g(t,x)$$

$$\frac{\partial p(x,t)}{\partial t} = D_p \frac{\partial^2 p(x,t)}{\partial x^2} \mp v_p \frac{\partial p(x,t)}{\partial x} - \frac{p(x,t)}{\tau_p} + g(t,x) \qquad (6.1)$$

where $\tau_n$ and $\tau_p$ are the excess carrier lifetimes, $D_n$, $D_{n1}$, $D_p$ and $D_{p1}$ are the diffusion coefficient of the electrons and holes outside and inside depletion region, respectively, $g(x,t)$ is the volume generation rate due to a light input, and $v_n$ and $v_p$ are the hole and electron drift velocities. In general, these velocities depend on the electric field. Since the photodiode is reversely biased, and the depletion region in poly diode without intrinsic layer is relatively small, there is a strong electric field inside the depletion region so drift velocities are maintained at their saturation values.

When the input light pulse is incident on the device, the generation rate $g(x,t)$ is:

$$g(x,t) = \Phi(1-R)[H(x) - H(x-L)]\frac{(1-e^{-\alpha K})}{K}\delta(t) \qquad (6.2)$$

where $\Phi$ is the incident light flux, $R$ is reflectivity of the surface, $K$ is the depth of the polysilicon layer, $l$ is the width of the polysilicon layer, $\alpha$ is absorption coefficient and $H$ and $\delta$ are Heaviside and Dirac pulses, respectively.

One way to solve drift-differential equations is first to simplify them by two substitutions. The substitution $n(x,t) = \exp(-t/\tau_n)N(x,t)$ is placed into the drift-diffusion equation (6.1), where $\tau_n$ is the electron recombination lifetime. This reduces (6.1) to:

$$\frac{\partial N(x,t)}{\partial t} = D_n \frac{\partial^2 N(x,t)}{\partial x^2} \pm v_n \frac{\partial N(x,t)}{\partial x} + g(t,x) \tag{6.3}$$

Then, substituting $\zeta = x \pm v_n t$ and $\theta = t$ into equation (6.2), the following partial differential equation is obtained:

$$\frac{\partial N(\xi,\theta)}{\partial \theta} = D_n \frac{\partial^2 N(\xi,\theta)}{\partial \zeta^2} + g(\zeta,\theta) \tag{6.4}$$

The above equation is a well-know equation of thermal conduction [9] and the final solution (after restoring the variables) is:

$$
\begin{aligned}
n(x,t) = \ & \Phi(1-R)e^{-\frac{t}{\tau_n}}\frac{(1-e^{-\alpha K})}{K}H(t) \\
& \times \frac{1}{2}\left[\mathrm{erf}\left(\frac{L-x\mp v_n t}{2\sqrt{D_n t}}\right) + \mathrm{erf}\left(\frac{x\pm v_n t}{2\sqrt{D_n t}}\right)\right]
\end{aligned}
\tag{6.5}
$$

A similar analytic expression follows for holes, by simply replacing $\pm v_n$ with $\mp v_p$ and $D_n$ with $D_p$.

The associated photocurrent $i_1(t)$ can be obtained by volume integration [6] of the conduction current density which consists of the photo-generated carriers moving over the graded depletion region, and dividing the result by the depletion region width $L$:

$$
\begin{aligned}
i_1(t) = \ & \frac{qW}{h\nu}(1-R)\frac{(1-e^{-\alpha K})}{K}\Phi H(t) \\
& \times \sum_{j=n,p} e^{-\frac{t}{\tau_j}}\left[E_1(t,v_j,D_j) + E_2(t,v_j,D_j)\right]
\end{aligned}
\tag{6.6}
$$

where $W$ is the width of the poly photodiode (see figure 6.1). The functions $E_1(t,v_j,D_j)$ and $E_2(t,v_j,D_j)$ are defined in terms of error functions and exponential functions, respectively:

$$E_1(t, v_j, D_j) = -(D_j + v_j^2 t)\mathrm{erf}\left(\frac{v_j t}{2\sqrt{D_j t}}\right)$$

$$-\frac{1}{2}\mathrm{erf}\left(\frac{L - v_j t}{2\sqrt{D_j t}}\right)[v_j^2 t - v_j L + D_j]$$

$$+\frac{1}{2}\mathrm{erf}\left(\frac{L + v_j t}{2\sqrt{D_j t}}\right)[v_j^2 t + v_j L + D_j]$$

$$E_2(t, v_j, D_j) = \frac{v_j\sqrt{D_j t}}{\pi}\left[\exp(-\frac{(L - v_j t)^2}{4D_j t})\right.$$

$$\left. + \exp(-\frac{(L + v_j t)^2}{4D_j t}) - 2\exp(-\frac{v_j t^2}{4D_j t})\right] \tag{6.7}$$

For the case where the diffusion inside the junction is negligible and excess carrier lifetime is longer than the carrier transit time (as is the case for CMOS poly-diodes without an intrinsic layer), the impulse response of the polysilicon diode can be simplified to:

$$i_1(t) = qW\Phi(1 - R)\frac{(1 - e^{-\alpha K})}{K}\delta(t)\sum_{j=n,p} v_j(L - v_j t)H(L - v_j t) \tag{6.8}$$

where $L$ is the length of the poly photodiode (see figure 6.1).

For polydiodes with a large intrinsic layer, the recombination lifetime is much shorter than the transit time and the impulse response is given in [6].

Because of the narrow depletion region ($<0.5$ $\mu$m), the diffusion length of the excess carriers is larger than the depletion region width and there are almost no carriers recombined in this region. Moreover, the excess carrier profile $n, p(x, t)$ is almost constant and the simplified formula for the drift frequency $f = 0.4v_s/L$ can be used, where $v_s$ is the saturation velocity of the excess carriers.

If the recombination process dominates the response of the polysilicon diode, one can take the recombination time much shorter than the transit time. The impulse current response of the lateral polysilicon diode can be then expressed as [9]:

$$i(t) = \frac{q}{h\nu L}(1 - R)[1 - \exp(-\alpha K)]\theta(t) \sum_{j=\mathrm{n,p}} v_j \exp(\frac{t}{\tau_j}) \qquad (6.9)$$

## 6.1.2   Diffusion current outside the depletion region

If the polysilicon photodiode is realized using two highly (inversely) doped regions without an intrinsic layer in between, the width of the depletion region is very small and the diffusion current outside this region will also contribute the overall photocurrent. This diffusion current is calculated in the n-region and p-region using the procedure similar to that explained in chapter 3. Here, the one-dimensional lateral diffusion equation is solved. Starting from the diffusion equations, the carrier profile is calculated using the boundary conditions shown in figure 6.3:

- the excess carriers concentration on the edge of depletion region is zero.

- the excess carriers concentration on the diffusion distance $L_j$, $j=n, p$ is zero.

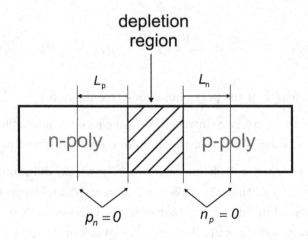

Figure 6.3: The boundary conditions for the diffusion current inside polysilicon diode.

From the carrier profile, the diffusion current can be calculated at the border of the depletion region:

$$i_2(t) = 4qWL\Phi\frac{1-e^{-\alpha K}}{K}\sum_m\sum_{j=n,p}\frac{L_j}{\tau_j}e^{-[(1+(2m-1)^2\pi^2]\frac{t}{\tau_j}} \qquad (6.10)$$

If the photodiode consists of $N$ n-p fingers, the total photocurrent $i_{tot}$ is directly proportional to the number of fingers, $i_{tot} = N \cdot i_2(t)$. Due to the exponential term in the (6.10), the speed of the diffusive response is mainly determined by the lifetime of the excess carriers. Noting that this lifetime is short (50 ps), the response speed of diffusion current component will be in hundreds of ps range. The overall photocurrent is the sum of drift and diffusion currents.

## 6.1.3   Frequency characterization of the polysilicon photodiode

The light sensitive part of a poly photodiode is only a small depletion region area plus the area outside this region within roughly one diffusion length of holes and electrons. This diffusion length is very small in comparison to that in monosilicon. The depth of the polysilicon in standard CMOS technology ($K$ in figure 6.1) is only about $0.2\mu$m and it also contributes to the poor responsivity of poly photodiodes on vertical incident light. According to this, a single polydiode would have very small active area and very low quantum efficiency (<1 % at $\lambda = 850$ nm). In order to increase the active photodiode area, a "jagged" polysilicon diode consisting of a number of polydiodes connected in parallel was realized (figure 6.4.). The overall active area is about 13 times larger than that in the single polydiode. This implies that the expected output signal is 22 dB larger than in a single polydiode. However, there are rounding effects at the many corners in the poly p-n structure that decrease this value.

The poly-diode is designed in a standard 0.18 µm CMOS technology, and the diode-layout is shown in figure 6.5. The measurements of the photocurrent showed that the actual photocurrent is 17 dB larger than the measured photocurrent of the single polysilicon photodiode.

The frequency response of the photocurrent is measured using an Agilent E4404E Spectrum Analyzer. The response of the polysilicon photodiode is measured from 1 MHz up to 6 GHz. For frequencies up to 1 GHz, the signal from the photodiode was amplified using a Minicircuits ZFL 1000LN 0.1-1000 MHz amplifier. For frequencies above, we used a 0.5-26.5 GHz Agilent 83017A Microwave system amplifier.

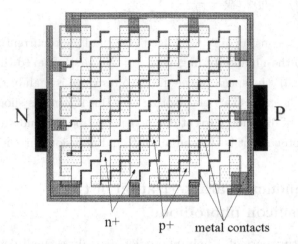

Figure 6.4: "Jagged" poly photodiode with an order of magnitude larger light sensitive area in comparison with a single poly photodiode. The lateral diode dimensions are $45 \times 45$ $\mu$m.

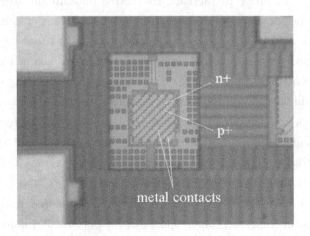

Figure 6.5: Layout of "jagged" poly photodiode designed to increase the overall light sensitive area.

The transmitter part consist of the 850 nm 10 Gb/s VCSEL and its driver amplifier. An HP 8665B frequency synthesizer was used as a modulating signal source up to 6 GHz. The signal was coupled into the photodiode using the multimode fiber with 50 μm core diameter.

The same setup is for calibration purposes used to measure a reference photodiode (Tektronix SA-42) response, which has according to specifications, 7 GHz -3 dB frequency. The response of the reference diode in the setup is presented in figure 6.6.

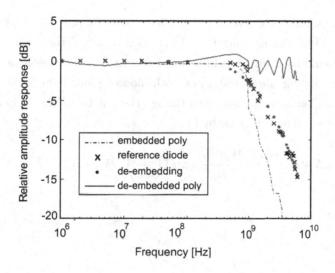

Figure 6.6: Frequency response of de-embedded polysilicon photodiode.

The polysilicon photodiode frequency characteristic was de-embedded, showing the solid curve in figure 6.6. The characteristics is almost flat up to 6 GHz, meaning that the measured bandwidth of the polydiode is even larger. The high intrinsic (physical) bandwidth is due to the short excess carrier lifetime (about 50 ps [2]), as described in section 5.7. The capacitance of the poly-diode using the 0.18 μm technology parameters is small ∼ 0.2 pF, which results in the large electrical bandwidth in our measurement setup.

## 6.2    Noise in polysilicon photodiodes

Large leakage current in polysilicon photodiodes for rather low values of reverse voltages (20µA for 1.5 V) causes a high noise in the photodiode which limits performance. For this reason, the following section presents a leakage current in a polysilicon photodiode.

### 6.2.1    Dark leakage current in the polysilicon diode

Figure 6.2 shows the diode reverse I-V characteristic of a polysilicon photodiode without light. The leakage current is large as a result of the grain-boundary trap-assisted band-to-band tunnelling and field-enhanced emission rate [5, 10]. Also, since the doping concentration of both diode regions is high the width of the depletion region is very small, even though the junction behaves as a graded one. The reverse current is given by [4]:

$$J_r = qN_t kT\pi \frac{\sigma v_{th} n_i}{2} \frac{W_d(V_R)}{L_g} \exp\left[\left(\frac{\alpha}{E_0}\right)^n (V_R + V_b)^{\frac{2n}{3}}\right] \tag{6.11}$$

where

$$\alpha = \left(\frac{9}{32}\frac{qa}{\epsilon}\right)^{\frac{1}{3}} \tag{6.12}$$

with $a$ [cm$^{-4}$] is the dopant concentration gradient and $V_R$ and $V_b$ are applied and built-in potentials voltages respectively. $E_0$ is the threshold electric field in the depletion region from which the emission amplification becomes significant (depending on the temperature and on the material), $v_{th}$ is thermal velocity often given as $v_{th} = \sqrt{3kT/m_{e,h}}$ where $m_{e,h}$ is the mass of the electron or hole, $n_i$ is intrinsic carrier concentration, $\sigma$ is an effective capture cross-section [cm$^2$], $W_d(V_R)$ is depletion region width [cm], $L_g$ is the grain size in polysilicon [cm], $N_t$ is the grain boundary trap density [cm$^{-2}$eV$^{-1}$] and $n$ is the exponential argument which generally varies between 0 and 1.5.

The above equation includes field enhancement of the emission rates of traps in the depletion region [11]. The value of $E_0$ depends also on the junction area. In our case we took the approximated value of $E_0 = 2 \cdot 10^5$ V/cm. H.C. de Graaf et. al. showed in [12] that the trap energy distribution $N_t$ is $U$-shaped with the broad minimum around mid-gap. For most purposes it can be approximated by a homogeneous distribution with $N_t = 3 - 5 \cdot 10^5$ cm$^{-2}$eV$^{-1}$. The capture-cross section for polysilicon is about $\sigma = 10^{-15}$cm$^2$, and the thermal velocity is about $1.2 \cdot 10^5$ m/sec.

According to both calculations and measurements of the reverse diode characteristic, it follows that if the reverse voltage value is higher than 0.7 V, the leakage current is higher than 500 nA. The high leakage current results in a high shot-noise that decreases the sensitivity of the polysilicon photodiode. For higher voltages (>1.5 V), the value of the leakage current can be even higher than the magnitude of the photocurrent, and this poly-diodes can be used only in high optical-power applications like detection of pulsed light signals and for trigger applications.

## 6.3 Time domain measurements

The characterization of the polysilicon photodiode is also performed in the time domain. Firstly, a picosecond laser with $\lambda = 650$ nm was used as a transmitter. The pulse width of the picosecond laser is 200 ps and the peak optical power is 1 mW (0 dBm). This rather large optical power was necessary due to the low quantum efficiency of poly-diode, which will be shown in section 5.10 of this chapter. The light was coupled from the laser to the poly-diode using multimode fiber with 50 μm core-diameter. The poly-diode was not packaged, and "on-chip" measurements were done using RF probes. The diode DC biasing of $V_R$=-0.5 V, was provided using a bias-tee. Larger (negative) voltages cause significant leakage currents ($> 0.5\mu$A), see figure 6.2.

The alignment of the fiber on the photodiode was done using micro-manipulators of the probe-station. By shining the light from the pulse-laser, the RF signal from the poly photodiode shown in figure 6.4 is measured first with an external amplifier with 750 Ω transimpedance; the result is shown in figure 6.7. The maximum measured output voltage is 1.2 mV, meaning that the maximum photocurrent is 1.6 μA. Since the maximum input optical power is 1 mW, the poly-diode responsivity as well as its quantum efficiency is clearly very low. The exact numbers are given in section 6.4.

The pulse width in figure 6.7 is about 1 ns which is larger than calculated in previous sections of this chapter. This is a measurement of *embedded* polydiode inside the resistances, capacitances and inductances of the bondpads, connec-

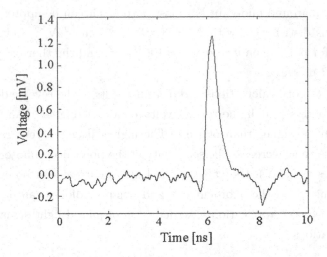

Figure 6.7: Transient response of poly photodiode on 200 ps input light pulse (transimpedance 750 $\Omega$, $\lambda = 650$ nm).

tors, and series resistances of the diode itself. In order to *de-embed* [6] the poly-diode we used the Tektronix SA-42 photodetector with 7 GHz-3 dB performance. The same TIA and coaxial cables are used for the measurements. The measured time response of this photodetector on 650 nm, is presented in figure 6.8.

In order to do the de-embedding i.e. calibrating out the measurement equipment, the equation above has to be solved for $R_{\mathrm{d}}(x)$. This is a complex deconvolution problem that can be only solved numerically [13]. The resulting response of the de-embedding polydiode is shown in figure 6.8. The estimated speed of the de-embedded polysilicon photodiode is at least as fast as a reference diode which has 7 GHz cutoff frequency.

Secondly, a picosecond laser with $\lambda = 400$ nm was used as a transmitter and the output signal from the polysilicon photodiode is presented in figure 6.9. The signal shape is similar to that shown in figure 6.8 with four times larger signal amplitude. This complies with the earlier (theoretical) findings reported in chapter 2 since the absorbtion coefficient of light in polysilicon is four time larger. Due to the fact that the speed of the polydiode at $\lambda=400$ nm is higher than that of the reference photodiode 6 GHz, it was impossible to accurately de-embed.

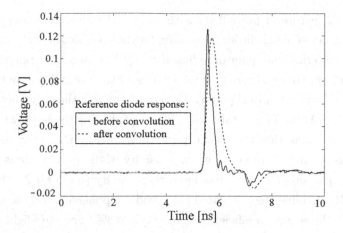

Figure 6.8: Transient response of the reference photodiode (7 GHz-3 dB, transimpedance 750 ohm, $\lambda = 650$ nm) and its convolution with the *difference* between embedded and de-embedded poly photodiode)

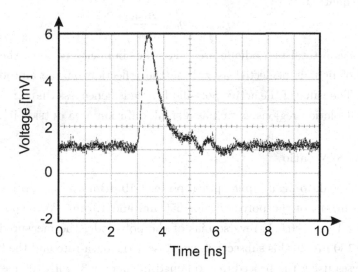

Figure 6.9: Transient response of embedded poly photodiode on 200 ps input light pulse (transimpedance 750 $\Omega$, $\lambda = 400$ nm).

## 6.4    Quantum efficiency and sensitivity

An important feature of polysilicon is that the light absorption depth is four times larger than in monosilicon. Therefore, for the same depth of the polysilicon and silicon material, the quantum efficiency (QE) is larger for polysilicon [2]. Previous sections showed that the photocurrent of the polysilicon photodiode is 1.6 µA for 1 mW input optical power. The responsivity of the poly photodiode is thus only 1.6 mA/W. The metal coverage area of the polydiode shown in figure 6.5 is 15 %, meaning that the optical power absorbed by the active area of the poly-diode is 8.5 mW. However, the responsivity of the poly-diode is still very low. Using equation (2.14) the maximum responsivity ($\eta$=1) for $\lambda$ = 650 nm is 0.52 A/W. By dividing the calculated poly-diode responsivity and the maximum responsivity, the quantum efficiency is only $\eta = 0.3\%$. For blue-light, $\lambda = 400$ nm, the photocurrent is 8 µA for 1 mW optical power; resulting in a responsivity of 8 mA/W. The maximum responsivity for $\lambda = 400$ nm is 0.32 A/W: the quantum efficiency is thus only $\eta = 2.5\%$.

Using a simplified formula for the maximum achievable quantum efficiency for both wavelengths $\eta_{\mathrm{max}} \sim 1 - e^{-\alpha K}$, the values are 21% and 97% respectively. The active (light sensitive) detector area $A_{\mathrm{eff}}$ can be estimated using the following equation:

$$\eta_{\mathrm{meas}} = \eta_{\mathrm{max}} \frac{A_{\mathrm{tot}}}{A_{\mathrm{eff}}} \qquad (6.13)$$

where $A_{\mathrm{tot}}$ is a total photodiode area. For the simplicity reasons, the bottom reflection of light is neglected as well as the reflection on the air/polysilicon interface. The value of the active poly-diode area is hence less than 2% implying very thin depletion regions as well as a small diffusion area outside it[1].

### BER and $S/N$ ratio

For 25 µW peak-to-peak input optical power (-19 dBm average optical power) the photocurrent of the polydiode for 650 nm and 1.6 mA/W responsivity is 40 nA. For $V_R = -0.5$ V reverse bias of the polydiode, the measured leakage current is 180 nA. In this subsection, we present the data-rate and the bit-error-rate analyses using the procedure explained in chapter 3, with the assumption that the subsequent TIA is noise-free. To achieve $S/N$=8 for BER= $10^{-12}$ the noise current is maximally 5 nA; the bandwidth of the polydiode for -19 dBm

---

[1]Multiplying the maximum quantum efficiency $\eta_{\mathrm{max}}$ with the calculated active poly-diode area 21% · 2% the poly-diode quantum efficiency for $\lambda = 650$ nm is about 0.4%.

input optical power is then limited to only 400 MHz. Taking noise from the TIA into account, the bandwidth is even lower than 400 MHz for BER=$10^{-12}$. Moreover, for larger bias voltages ($V_R$ >-1 V), the leakage current of the poly-diode increases dramatically as shown in figure 6.2. This large leakage limits the poly-diode bandwidth in the low MHz range.

Improvement of the quantum efficiency in poly-diodes can typically be done using two methods. First, light reflectors can be used which creates a resonant-cavity photodiode [7]. This is however not possible in standard CMOS technology. The second method is to design a PIN poly photodiode [6], which includes non-doped polysilicon layer, which is also not available in standard CMOS technology.

## 6.5   Conclusion

This chapter described a lateral polysilicon photodiode in standard 0.18 µm CMOS technology. The analytical calculations, and the measurements in the frequency and the time domain showed that polysilicon photodiode has a very large bandwidth: $f_{3dB}$ > 6 GHz. Due to the small excess carrier lifetime, the slow diffusion limitation on the intrinsic (physical) polydiode bandwidth is negligible. The electrical bandwidth limitation is also minimal: the small diode parasitic capacitance is proportional to the low depth of the polysilicon layer. The big advantage of polydiode is that the parasitic capacitance towards the substrate is also very low because of the thick field oxide layer in comparison with the conventional thin gate oxide.

The disadvantage of the polydiode in standard CMOS technology is the low quantum efficiency ($\leq$2.5 %). This is because of the very small light sensitive area: the *width* of the depletion region is small because of the high doping concentrations of the n-region and p-region. The *depth* of the poly-diode is limited by the technology. The "out of junctions" active diode area is also small due to the small diffusion lengths of the excess carriers. These diffusion lengths are determined by the short carrier lifetime (50 ps).

There are a few ways to improve this low quantum efficiency, that are however not feasible in standard CMOS: adding light reflectors resulting to make a resonant-cavity photodiode and using lightly doped poly areas to make a poly PIN photodiode.

# Bibliography

[1] S.Radovanovic, A.J.Annema and B.Nauta, *"High-speed lateral polysilicon photodiode in standard CMOS"*, IN pROC. ESSDERC 2003, Estoril, Portugal, pp.521-524.

[2] Kamins T.: *"Polycrystalline silicon for integrated circuits and displays"*, Boston : Kluwer Academic Publishers, 2nd edition, p. 240, 1998.

[3] McKelvey, J. P.: *"Solid-State and Semiconductor Physics"*, New York: Harper& Row, pp. 340, 1966.

[4] A. Aziz, O. Bonnaud, H. Lhermite and F. Raoult: "Lateral polysilicon pn diode: Current-voltage characteristics simulation between 200K and 400K using a numerical approach", *IEEE Transactions on Electron Devices*, vol. 41, pp. 204-211, 1994.

[5] Manoliu and T. Kamis, "P-N junction in polycrystalline-silicon films", *Solid-State Electronics*, Vol. 15, pp. 1103-1106, 1972.

[6] Kim, D. M., Lee, J. W., Dousluoglu, T., Solanki, R. and Qian, F: "High-speed lateral polysilicon photodiodes", *Semiconductor Sci. Technology*, vol. 9, pp. 1276-1278, 1994.

[7] Diaz, D.C., Scho, C.L., Qi, J, Campbell, J.C.: "High-speed Polysilicon Resonant-Cavity Photodiode with $SiSO_2 - Si$ Bragg reflector", *Photonics Technology Letters*, vol. 9, no.6, pp. 806-808, June, 1997.

[8] Plummer J., Deal M., Griffin P.: *"Silicon VLSI technology; fundamentals, practice and modelling"*, Prentice Hall, p. 560, 2000.

[9] Lee, J.W., Kim., D. M.: "Analytic time domain characterization of $p$-$i$-$n$ photodiodes: effects of drift, diffusion, recombination, and absorption", *Journal of Applied Physics*, vol. 6, pp. 2950-2958, March 1992.

[10] M. Dutoit and F. Sollberger: "Lateral Polysilicon p-n Diodes", *Solid-State Science and Technology*, vol.125, No.10, pp1648-1651, October 1978.

[11] D. W. Greve, P. Potyraj and A. Guzman: "Field-enhanced emission and capture in polysilicon pn junctions", *Solid-State Electronics*, vol. 28, pp. 1255-1261, 1985.

[12] H.C. de Graaf, M. Huybers: "Grain-boundary states and the characteristics of lateral polysilicon pn junctions", *Solid-State Electronics*, vol. 25, pp. 67-71, 1982.

[13] P. C. Hansen: *"Deconvolution and regularization with Toeplitz matrices"*, Numerical Algorithms, vol. 29, 2002, pp. 323-378.

# CHAPTER 7

# CMOS photodiodes: generalized

*This chapter presents analyses of the frequency behavior of photodiode in standard CMOS for the whole wavelength range: 400 nm<λ<850 nm. Independent of CMOS technology, for all wavelengths for which most of the light is absorbed at depths smaller than that of the most shallow junction, shorter wavelengths result in a lower bandwidth. This bandwidth is however still in the hundreds of MHz range.*

*For wavelengths for which the 1/e-absorption depth is much larger than the deepest junction depth, the substrate current dominates the total response. Here shorter wavelengths results in a larger photodiode bandwidth. This chapter generalizes the findings and solutions in chapters 3, 4 and 5 to any sensible wavelength range and to different CMOS technologies.*

## 7.1  Introduction

The major effect of different wavelengths on the physical behavior of the photodiode is that the penetration depth of the light is a strong function of the wavelength, see e.g. [1]. For example at 850 nm, the depth at which 50% of the light is absorbed is about 9 $\mu$m, for $\lambda$=600 nm this depth is 1.8 $\mu$m, and down to only 0.7 $\mu$m at $\lambda$=500 nm. Therefore, at shorter wavelengths the light is absorbed in the upper parts of the photodiode which typically results in faster response of CMOS photodiodes.

This chapter analyzes the frequency response of CMOS photodiode in general for the wavelength range 400nm<$\lambda$<850 nm. For this purpose, a device-layer/p-substrate photodiode is used, as shown in figure 7.1. Device layers are those that make an active semiconductor device (diodes in this case), like nwell, n+ and p+ layers and this layer is assumed as the top layer of the CMOS photodiode. The influence of different photodiode regions on the total photocurrent will be analyzed in detail. The depth of the device-layer is in the range from 4 µm to 0.05 µm in nowadays and upcoming CMOS generations.

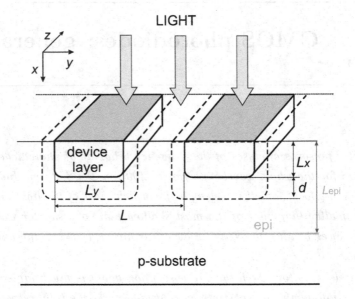

Figure 7.1: A general photodiode in standard CMOS technology.

In the second part of the chapter, the findings and solutions in chapters 3 and 4 are generalized to any wavelength and CMOS generation.

## 7.2 Generalization of CMOS photodiodes

Equation (3.1) can be rewritten as a function of CMOS technology-scale. This can be done by using both the depth of the device layer and the wavelengths related to the light penetration depth. In general, all photocurrent components defined in (3.1) are also dependent on the dimensions of the corresponding diode regions. However, to stress the importance of the technology and wavelength on the total diode response, these two parameters are explicitly included in (7.1). For the case where the device-layer depth is lower than the light absorption depth $1/\alpha$, all diode layers contribute to the overall photocurrent:

$$
\begin{aligned}
I_{\text{tot}}(L_x, \lambda, s) = & \; I_{\text{drift}}(L_x, \lambda) \frac{1}{\sqrt{1 + \dfrac{s}{s_{\text{drift}}(L_x, \lambda)}}} \\
& + \sum_k I_{\text{diff}_k}(L_x, \lambda) \frac{1}{\sqrt{1 + \dfrac{s}{s_{\text{diff}_k}(L_x, \lambda)}}}
\end{aligned}
\tag{7.1}
$$

where $I_{\text{diff}_k}(L_x, \lambda)$ and $I_{\text{drift}}(L_x, \lambda)$ are the amplitudes of the diffusion and the drift regions of photodiode and $s_{\text{diff}_k}$ and $s_{\text{drift}}$ are the poles of the diffusion currents and drift current, respectively. Chapter 3 showed that diffusion and drift processes in general have a low roll-off ($\sim$10 dB/decade). For this reason, the individual current components are approximated here as square-root pole functions; this also provides easier understanding of the total photodiode behavior.

For device-layer depths larger than the light absorption depth $1/\alpha$, the total photocurrent is dominated by the diffusion current and the drift currents in the depletion regions. The diffusion current is given in (3.8); the drift current in (3.17). The following sections of this chapter discuss the photocurrent components (7.1) and the total photocurrent as a function of CMOS technology and input wavelength. The amplitude of the photocurrent components will be normalized with the amplitude of the total photocurrent $I_{\text{tot}}$.

## 7.3   Device layer - photocurrent amplitude

The first photocurrent component in (7.1) is the device-layer diffusion current. This section investigates in particular *maximum amplitude* of this diffusion current, as a function of the CMOS technology and wavelength. For simplicity reasons the quantum efficiency is assumed to be maximal ($\eta$=100 %).

The amplitude of the device-layer photocurrent is calculated following the procedure given in section 3.2.1. That section analyzed the nwell diffusion current as a function of frequency. Replacing the hole diffusion length $L_p$ in equation (3.8) with general diffusion length $L_{\mathrm{gen}}$, the maximum amplitude of the diffusion current in the device-layer (nwell, n+ or p+) is obtained for $s\tau_p \ll 1$:

$$
\begin{aligned}
I_{\mathrm{diff}_{DL}}(L_x, \lambda) = \;& 32\Phi_0 \frac{eL_{\mathrm{gen}}^2 \alpha}{l\pi^2} \sum_{n=1}^{\infty}\sum_{m=1}^{\infty} \frac{(2n-1)\pi e^{-\alpha L_x} + (-1)^{\frac{(2n-1)-1}{2}}\alpha L_x}{4\alpha^2 L_x^2 + (2n-1)^2\pi^2} \\[2mm]
& \times \; \frac{\dfrac{2L_{x1}}{L_y}\dfrac{1}{2n-1} + \dfrac{L_y}{2L_x}\dfrac{2n-1}{(2m-1)^2}}{\dfrac{(2n-1)^2\pi^2 L_{\mathrm{gen}}^2}{4L_x^2} + \dfrac{(2m-1)^2\pi^2 L_{\mathrm{gen}}^2}{L_y^2} + 1}
\end{aligned}
\tag{7.2}
$$

where $L_x$ and $L_y$ are the width and length of the device-layer (shown in figure 7.1) and $l$ is the distance between subsequent device-layers. The diffusion length $L_{\mathrm{gen}}$ mainly depends on a doping concentration of the layer [1].

The calculated device-layer current $I_{\mathrm{diff}_{DL}}(L_x, \lambda)$ is shown in figure 7.2. For wavelengths for which the 1/e-absorption depth is larger than the device-layer depth $L_x$, *larger* wavelength results in lower photocurrent. On the other side, for wavelengths for which the light is almost completely absorbed in the device-layer, the photocurrent corresponds to the diode responsivity.

### 7.3.1   Device layer - photocurrent bandwidth

The device-layer current is the diffusion current of minority carriers. The bandwidth of the diffusion current depends on the wavelength and CMOS technology. For easier understanding of this dependence, first the light absorption inside the device-layer as a function of depth as shown in figure 7.3. For all wavelengths with the 1/e-absorption depth *larger* than device-layer depth $L_x$ ($L_x \ll 1/\alpha_1, 1/\alpha_2$), the light absorption inside the device layer is almost constant with distance:

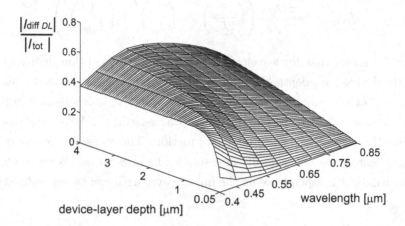

Figure 7.2: The normalized photocurrent of device-layer region versus layer depth (technology dependent) and input wavelength, for device-layer/p--substrate photodiode. The device-layer photocurrent is normalized with a total diode photocurrent.

$$\exp(-\alpha_1 x) \approx \exp(-\alpha_2 x) \approx Const. \tag{7.3}$$

Therefore, the excess carrier concentration gradient towards the junctions can be taken to be independent of input wavelength, [2]. Chapter 3 showed that device-layer bandwidth is proportional to the excess carrier concentration gradient. This gradient is constant, and the bandwidth is wavelength-independent; with (3.8) it follows that:

$$f_{3dB} \approx \frac{\pi D_{gen}}{2} \left( \left( \frac{1}{2L_x} \right)^2 + \left( \frac{1}{L_y} \right)^2 + \left( \frac{1}{L_{gen}} \right)^2 \right) \tag{7.4}$$

where $D_{gen}$ is the general diffusion constant of the excess carriers. This -3dB frequency can easily be translated into a cut-off frequency; for a slope of $-10\mathrm{dB}/\text{decade}$ this yields

$$f_{\text{cut-off}} \approx \frac{\pi D_{\text{gen}}}{2\sqrt{3}} \left( \left(\frac{1}{2L_x}\right)^2 + \left(\frac{1}{L_y}\right)^2 + \left(\frac{1}{L_{\text{gen}}}\right)^2 \right) \tag{7.5}$$

Figure 7.3 shows that for wavelengths with the $1/e$-absorption depth *smaller* than the device-layer depth $(L_x)$, the total photocurrent is generated mainly in the layer. For very small wavelengths ,e.g. for $\lambda_5$, the light is almost completely absorbed inside the layer. The lower the wavelength, the larger the distance between the generated carriers and the junction. The excess carrier concentration gradient drops which results in a smaller bandwidth, as shown in chapter 3. The bandwidth dependence on the input wavelength can be generalized from (3.8):

$$f_{\text{3dB}} \approx \left(\frac{\lambda}{\lambda_{1/e}}\right)^{\frac{1}{3}} \frac{\pi D_{\text{gen}}}{2} \left( \left(\frac{1}{2L_x}\right)^2 + \left(\frac{1}{L_y}\right)^2 + \left(\frac{1}{L_{\text{gen}}}\right)^2 \right) \tag{7.6}$$

$$f_{\text{cut-off}} \approx \left(\frac{\lambda}{\lambda_{1/e}}\right)^{\frac{1}{3}} \frac{\pi D_{\text{gen}}}{2\sqrt{3}} \left( \left(\frac{1}{2L_x}\right)^2 + \left(\frac{1}{L_y}\right)^2 + \left(\frac{1}{L_{\text{gen}}}\right)^2 \right) \tag{7.7}$$

For very small device-layer depths $(< 0.2\mu\mathrm{m})$, the doping concentration of the layer is usually high. As a result, the diffusion constant of the minority carriers is smaller [1]. The device-layer bandwidth, presented in (7.6) is proportional to the ratio $D_{\text{gen}}/L_x$ and $D_{\text{gen}}/L_y$. This bandwidth increases with the downscaling of the device-layer: the diffusion constant $D_{\text{gen}}$ decreases "slowly" in comparison with the downscaling of $L_x$ and $L_y$.

For wavelengths for which the light is almost completely absorbed in the device-layer, the bandwidth and the photocurrent are proportional in nature. The lower the wavelength, the lower the photocurrent (due to the lower responsivity, see chapter 2), and the lower the device-layer bandwidth (since the majority of the carriers are generated further from the bottom junction).

## 7.3.2   Substrate current-photocurrent amplitude

The largest part of the photodiode in standard CMOS technology is its substrate. Chapter 3 described the substrate current response for $\lambda = 850$ nm; the light penetration depth, for this wavelength is about 30 µm (99% of the light

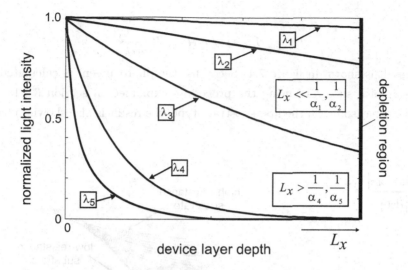

Figure 7.3: Normalized light intensity inside the device-layer region for different wavelengths: $\lambda_5 < \lambda_4 < \lambda_3 < \lambda_2 < \lambda_1$.

absorbed). Therefore, for all recent and future CMOS technologies (feature-size <0.5 µm) the substrate current dominates on the overall photocurrent. Chapter 3 analyzed two types of p substrate: high-resistance and low-resistance substrates. It was shown that the photodiode bandwidth is larger for low-resistance substrate: the recombination rate in the heavily doped substrate if high which effectively kills carriers generated deep in the substrate. The effect of this is a high bandwidth at the cost of some responsivity.

Chapter 5 described photodiode behavior on $\lambda = 400$ nm. Most of the carriers are generated close to the photodiode surface i.e. close to diode junctions. These carriers diffuse faster towards junctions and the bandwidth is in hundreds of MHz range. The influence of the substrate current on the overall photocurrent is negligible.

This section presents substrate current response of the photodiode in CMOS technology for the whole wavelength sensitivity range. Firstly, amplitude of the substrate photocurrent $I_{\text{diff}_{\text{subs}}}(L_x, \lambda)$ from equation (7.1) is calculated for *high-resistance* substrate, using the resulting relation between depth, wavelength and current (3.16):

$$I_{\mathrm{diff}_{\mathrm{subs}}}(L_x, \lambda) \cong e\alpha L_n e^{-\alpha L_x} \frac{1}{\alpha L_n} \tag{7.8}$$

The result is shown in figure 7.4. Secondly, the photocurrent is calculated for *low-resistance* substrate using the procedure explained in section 3.2.1. For easier comparison with the first substrate type, the result is also shown in figure 7.4.

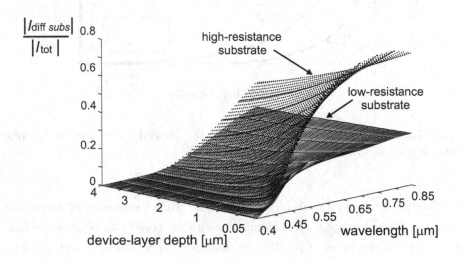

Figure 7.4: Normalized photocurrent of the substrate versus device-layer depth (technology dependent) and input wavelength, for photodiode in standard CMOS. The depth of the $L_{\mathrm{epi}}$ is assumed to be three times larger than the device-layer depth. The substrate current is normalized with the total diode current.

For wavelengths for which the absorbtion depth is smaller or equal to the depth of the epi-layer $L_{\mathrm{epi}}$ (see figure 7.1) in low resistance-substrate, the substrate photocurrent is comparable for both substrate types due to the similar "active" diode layers.

## 7.3.3   Substrate current-photocurrent bandwidth

Chapter 3 showed that the substrate current bandwidth does not depend on the device-layer depth: it is independent of CMOS technology. This can be explained using the fact that a shifted exponential curve equals a scaled expo-

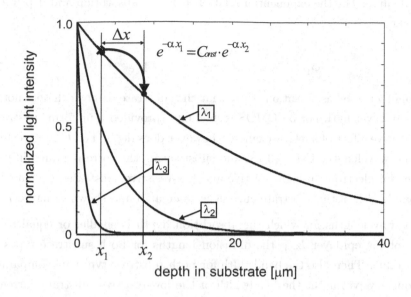

Figure 7.5: Normalized light intensity inside the substrate for different wavelengths: $\lambda_3 < \lambda_2 < \lambda_1$.

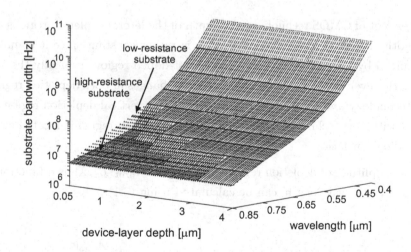

Figure 7.6: The bandwidth of substrate diffusion current versus device-layer depth (technology dependent) and input wavelength.

nential curve. For the exponential relation between absorbtion and depth for a constant $\lambda$:

$$\Phi_1 e^{-\alpha x_2} = \Phi_1 e^{-\alpha \Delta x} e^{-\alpha x_1} = \Phi_2 e^{-\alpha x_1} \tag{7.9}$$

Combining these basic relations show that the substrate bandwidth does not depend on wavelength nor on CMOS technology (provided a uniform substrate). The relative effect of substrate current however does depend on CMOS technology and wavelength. Using (3.15) and substituting the corresponding diffusion lengths for electrons in high-resistive and low-resistive substrates $L_{n1}$ and $L_{n2}$, the bandwidth of the substrate current is calculated and shown in figure 7.6.

For wavelengths for which the absorbtion depth is smaller or equal to the depth of the epi-layer $L_{epi}$, the diffusion lengths for both substrate types are comparable. Then also the bandwidth for both substrate types are comparable. For longer wavelengths the bandwidth of the low-resistive substrate current is higher.

### 7.3.4   Depletion region current

Independent of CMOS technology, the depth of the lateral depletion regions and the width of the vertical depletion regions is rather constant since it is mainly determined by the concentration of the lowest-doped region: typically the substrate. However, the absorbed amount of light changes in both depletion regions with technology and wavelength. The depth of the vertical depletion region decreases with new technologies, and the lateral depletion region is located closer to the diode surface.

The amplitude of depletion region photocurrent $I_{drift}(L_x, \lambda)$ as a function of technology and wavelength, can be calculated using (6.1):

$$\begin{aligned} I_{drift}(L_x, \lambda) &= \Phi e (e^{-\alpha L_x} - e^{-\alpha(L_x + d)}) \frac{A_{total}}{A_{eff_{lat}}} \\ &+ \Phi e (1 - e^{-\alpha L_x}) \frac{A_{total}}{A_{eff_{ver}}} \end{aligned} \tag{7.10}$$

The result is shown in figure 7.7.

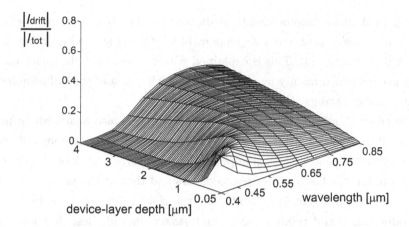

Figure 7.7: Normalized photocurrent of the depletion region versus device-layer depth (technology dependent) and input wavelength. A depletion region current is normalized with the total photocurrent.

### 7.3.5 Depletion region - photocurrent bandwidth

Chapter 3 described the transient time of excess holes and electrons in the depletion region. The corresponding - 3dB frequencies of hole and electron currents are presented too. Because of the limited biasing voltages in standard CMOS processes and having the depletion region width determined by doping concentration of the device-layer and the substrate, the electric field in the depletion region is not high enough for carriers to reach their saturation velocities. Instead, they travel with a lower speed determined by the position-dependent electric field.

Using (3.24) and (3.25) the average transient times of holes and electrons as well as the bandwidth of the current can be calculated for different technologies and wavelengths. As a result of scaling and its associated lower voltages and somewhat higher dope levels, the bandwidth, for *constant wavelength*, is *lower with downscaling* of the technology[1].

### 7.3.6 Total photocurrent

Previous sections analyzed the photocurrent components of photodiode in CMOS technology for the whole wavelength sensitivity range. The sum of all compo-

---

[1]the depth of depletion region stays rather constant but the electric field decreases due to the lower bias voltages.

nents gives the total photocurrent, as shown in (7.1). *Maximum* photocurrent for any wavelength is almost independent of CMOS structure i.e. independent of CMOS technology [4]. This is explained with the fact that the quantum efficiency is determined mainly by absorption coefficient $\alpha$ and the diffusion length of the minority carriers [1].

The effect of going to newer CMOS technologies is also analyzed: in newer (bulk-CMOS) technologies the device dimensions shrink. At constant wavelength of the incident light, this implies that relatively much less photocurrent is generated in the fast parts of photo diodes and that at the same time somewhat more photocurrent is generated in the slow substrate. For high-ohmic substrates this would result in somewhat slower photodiodes. For low-ohmic substrates (and assuming that the epi-layer thickness also shrinks) the photodiode will be faster but will also have a somewhat lower responsivity.

For the input wavelengths 650 nm$< \lambda <$ 850 nm, independent on the CMOS technology, photodiode bandwidth is dependent on both the technology and the wavelength as shown in figure 7.8.

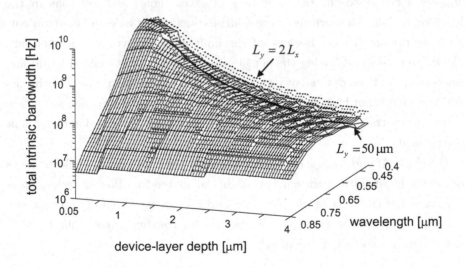

Figure 7.8: The total intrinsic bandwidth of general device-layer/p-substrate photodiode in standard CMOS versus device-layer depth and input wavelength.

# 7.4  Analog equalization

Chapter 4 showed that using an analog equalizer, slow photodiodes can be used in high-speed applications. The equalizer then compensates (in magnitude and phase) the low "roll-off" in the intrinsic diode response using a low "roll-up". Thanks to the low roll-off property this equalization is inherently robust against spread while robustness over temperature is quite good and can be improved by a simple feed-forward control loop.

Because the equalizer approach work very well only for low roll-offs, it can sensibly be applied from the cut-off frequency of the photodiode up to the frequency where the roll-off is high. In this context "high" is more than roughly 10dB/decade. Note that this upper limit in sensible equalization frequency can be due to:

- the electrical bandwidth, usually having a roll-off of -20dB/decade. This electrical bandwidth can be increased at the cost of power consumption.

- drift currents in the depletion layers, typically at a roll-off of -10dB/decade. For nowadays CMOS the cut-off frequency of drift currents is about 10GHz. It can be increased with e.g. higher voltages and more higher doped junctions.

- diffusion currents, each individual component having a roll-off of about -10dB/decade.

It can be concluded that inherently robust analog equalization can be done up to frequencies of about 10GHz. In that case a suitable pre-amplifier is required to get a sufficiently high electrical bandwidth. At the same time medium of short wavelengths must be used to get a sufficiently high cut-off frequency of the diffusion current components.

## 7.5   Summary and Conclusions

The intrinsic frequency characteristics of a CMOS photodiode depends on both the *wavelength* and the *technology*. Maximum diode bandwidth without equalization is achieved using:

- *short wavelength*, e.g. $\lambda = 400$ nm. The intrinsic bandwidth then up to about 8 GHz. This bandwidth is limited by the diffusion bandwidth.

- a *technology* for which the thickness of the device-layer is about the absorption depth of light, e.g. $1/\alpha_{400}$.

For other wavelengths and technologies, it is possible to achieve bandwidth enhancement using the analog equalization method explained in chapter 4. This bandwidth improvement is however dependent on wavelength, technology and device-layer width.

For device-layer depths larger than $1/e$ absorbtion depth of light $\alpha L_x > 1$, the roll-off in the intrinsic diode characteristics is high ($<15$ dB/decade). Due to this high roll-off the feed-forward equalization introduced in chapter 4 can increase the bit rate by just a small factor. For high roll-off figures the approach in chapter 4 is not inherently robust against e.g. spread.

For device-layer depths smaller than absorbtion depth of light $\alpha L_x < 1$, the roll-off in the intrinsic diode characteristics is $< 10$ dB/decade. The maximum equalization frequency is then limited by the drift bandwidth. The feed-forward analog equalization can be applied to achieve higher data rates. The usefull equalization range is however limited by robustness issues.

According to analyses in this section, maximum bandwidth improvement can be achieved using a single photodiode structure. Note that an associated advantage is that this both simplifies the layout and maximizes light-sensitive area.

# Bibliography

[1] S. M. Sze: *"Physics of semiconductor devices"*, New-York: Wiley-Interscience, 2-nd edition, 1981, p. 81.

[2] D. Coppée, H. J. Stiens, R. A. Vounckx, M. Kuijk: "Calculation of the current response of the spatially modulated light CMOS detectors", *IEEE Transactions Electron Devices*, vol. 48, No. 9, 2001, pp. 1892-1902.

[3] S. Alexander: *"Optical communication receiver design"*, SPIE Optical engineering press, 1997.

[4] I. Brouk, and Y. Nemirovsky: "Dimensional effect in CMOS photodiodes", *Solid State Electronics*, vol. 46, 2002, pp. 19-28.

# CHAPTER 8

Conclusions

## 8.1 Conclusions

In future communication systems for short distances (e.g. for chip-to-chip, board-to-board), optical interconnect may become important since straightforward electrical connections suffers from poor impedance matching, crosstalk and significant Electro-Magnetic noise which all degrade the system performance. Short distance communication channels are not shared by multiple users and as a result cost aspects are important for these non-shared channels. Because of this, existing solutions in long-haul communications cannot be used. To enable cost-effective implementation of optical short-distance interconnect, apart from low cost lasers and plastic fibers, standard CMOS technology for the electronics should be used,

For $\lambda=850$ nm, which is a typical wavelength used for short-haul communication, the bandwidth of the photodiodes in standard CMOS is in the low MHz range, which is *the* main limiting factor for the Gb/s optical detection. An in-depth analysis on the bandwidth of CMOS photodiodes for $\lambda = 850$nm is presented in chapter 3. A common feature of the frequency characteristics of the analyzed photodiodes is the low roll-off ($<5$ dB/decade) in the frequency range up to a few GHz. This stems from the fact that total photocurrent is the sum of the diffusion and drift currents having low roll-offs (10 dB/decade).

The low roll-off property of the photodiodes response enables the efficient use of a relatively simple analog equalizer to compensate it; this is the topic of chapter 4. In this way standard CMOS photodiodes can be used for high bitrates without sacrificing responsivity. The required equalization characteristic is the complement of the photodiode response: it has a low roll-up characteristic. Using this approach, 3 Gb/s data-rate is achieved with BER$< 10^{-11}$ and average input optical power of $P_{in}=25$ $\mu$W (-19 dBm). This is over half an order of magnitude speed-increase in comparison with state-of-the-art photodetectors in CMOS. The proposed pre-amplifier with the analog equalizer is inherently robust against spread and temperature variations.

Although the presented design is for $\lambda=850$ nm, the approach can be generalized to any wavelengths in the range from $\lambda=500$ nm to 850nm, see chapters 5 and 7. For even shorter wavelength equalization may not be required to obtain high data rates due to the low penetration depth at short wavelengths. The equalization is inherently robust against spread due to the low roll-off of the photodiode intrinsic response; no adaptive equalization is required. Changes up to e.g. 30% in temperature or in equalizer parameters result a modest performance degradation in terms of sensitivity and bitrate.

In chapter 6 we presented also a lateral polysilicon photodiode that has a frequency bandwidth far in the GHz range: the measured bandwidth of the poly photodiode was 6 GHz, which figure was limited by the measurement equipment. However, the quantum efficiency of poly-diodes is very low ($< 8$ %) due to the very small diode active area: the depletion region is very small due to the lack of intrinsic layer while the diode depth is limited by the technology.